SPARKNOTES
Power Tactics
FOR THE NEW SAT

THE MATH SECTION
GEOMETRY

SPARK NOTES

A DIVISION OF BARNES & NOBLE PUBLISHING

Copyright © 2005 by Spark Educational Publishing

All rights reserved. No part of this book may be used or reproduced in any manner whatsoever without the written permission of the Publisher.

SPARKNOTES is a registered trademark of SparkNotes LLC

Spark Educational Publishing
A Division of Barnes & Noble Publishing
120 Fifth Avenue
New York, NY 10011

ISBN 1-4114-0278-2

Please submit changes or report errors to *www.sparknotes.com/errors*

Printed and bound in Canada.

SAT is the registered trademark of the College Entrance Examination Board, which was not involved in the production of, and does not endorse, this product.

Written by Drew Johnson

CONTENTS

Introduction 4
About the New SAT 5
A User's Guide 8

The Power Tactics 9

Anatomy of SAT Geometry 10
Essential Concepts 15
Essential Strategies 35
Test-Taking Strategies 51
The 10 Most Common Mistakes 55
Conclusion 56

The Practice Sets 59

Practice Set 1 60
Answers & Explanations 62
Practice Set 2 67
Answers & Explanations 69
Practice Set 3 73
Answers & Explanations 76
Practice Set 4 80
Answers & Explanations 83
Practice Set 5 87
Answers & Explanations 88

INTRODUCTION

Truly effective SAT preparation doesn't need to be painful or time-consuming. SparkNotes' *Power Tactics for the New SAT* is proof that powerful test preparation can be streamlined so that you study only what you need. Instead of toiling away through a 700-page book or an expensive six-week course, you can choose the *Power Tactics* book that gets you where you want to be a lot sooner.

Perhaps you're Kid Math, the fastest number-slinger this side of the Mississippi, but a bit of a bumbler when it comes to words. Or maybe you've got the verbal parts down but can't seem to manage algebraic functions. *Power Tactics for the New SAT* provides an extremely focused review of every component on the new SAT, so you can design your own program of study.

If you're not exactly sure where you fall short, log on to **testprep.sparknotes.com/powertactics** and take our free diagnostic SAT test. This test will pinpoint your weaknesses and reveal exactly where to focus.

Since you're holding this book in your hands, it's pretty likely that SAT geometry is giving you trouble. You've made the right decision because in a few short hours, you will have mastered this part of the exam. No sweat, no major investment of time or money, no problem.

So, let's not waste any time: go forth and conquer SAT geometry so you can get on with the *better parts* of your life!

ABOUT THE NEW SAT

THE OLD

The SAT, first administered in 1926, has undergone a thorough restructuring. For the last ten years, the SAT consisted of two sections: Verbal and Math. The Verbal section contained Analogies, Sentence Completions, and Critical Reading passages and questions. The Math section tested arithmetic, algebra, and geometry, as well as some probability, statistics, and data interpretation.

You received one point for each correct answer. For most questions, a quarter of a point was deducted for each incorrect answer. This was called the "wrong-answer penalty," which was designed to neutralize random guessing. If you simply filled in the bubble sheet at random, you'd likely get one-fifth of the items correct, given that each item has five answer choices (excluding student-produced–response items). You'd also get four-fifths of the items wrong, losing $4 \times 1/4$, or 1 point for the four incorrectly answered items. Every time you determined an answer choice was wrong, you'd improve your odds by beating the wrong-answer penalty. The net number of points (less wrong-answer penalties) was called the "raw score."

Raw score = # of correct answers − ($1/4$ × # of wrong answers)

That score was then converted to the familiar 200–800 "scaled score."

THE NEW

For 2005, the SAT added a Writing section and an essay, changed the name of *Verbal* to *Critical Reading*, and added algebra II content to the Math section. The following chart compares the old SAT with the new SAT:

ABOUT THE NEW SAT

Old SAT	New SAT
Verbal	**Critical Reading**
Analogies	*Eliminated*
Sentence Completions	Sentence Completions
Long Reading Passages	Long Reading Passages
Paired Reading Passages	Paired Reading Passages
	Short Reading Passages
Math—Question Types	
Multiple Choice	Multiple Choice
Quantitative Comparisons	*Eliminated*
Student-produced Responses	Student-produced Responses
Math—Content Areas	
Numbers & Operations	Numbers & Operations
Algebra I	Algebra I
	Algebra II
Geometry	Geometry
Data Analysis, Statistics & Probability	Data Analysis, Statistics & Probability
	Writing
	Identifying Sentence Errors
	Improving Sentences
	Improving Paragraphs
	Essay
Total Time: 3 hours	*Total Time*: 3 hours, 45 minutes
Maximum Scaled Score: 1600	*Maximum Scaled Score*: 2400 Separate Essay Score (2–12)

The scoring for the test is the same, except that the Writing section provides a third 200–800 scaled score, and there is now a separate essay score. The wrong-answer penalty is still in effect.

| ABOUT THE NEW SAT |

NEW PACKAGE, OLD PRODUCT

While the test has changed for test-*takers*, it has not changed all that much from the test-*maker*'s point of view. The Educational Testing Service (ETS) is a not-for-profit institute that creates the SAT for The College Board. Test creation is not as simple a task as you might think. Any standardized test question has to go through a rigorous series of editorial reviews and statistical studies before it can be released to the public. In fact, that's why the old SAT featured a seventh, unscored, "experimental" section: new questions were introduced and tested out in these sections. ETS "feeds" potential questions to its test-takers to measure the level of difficulty. Given the complex and lengthy process of developing new questions, it would be impossible for ETS to introduce *totally* new question types or make major changes to the existing question types.

Now that you know these facts, the "new" SAT will start to make more sense. The changes were neither random nor unexpected. Actually, the only truly *new* question type on the SAT is the short reading passages followed by a couple of questions. However, the skills tested and strategies required are virtually identical to the tried-and-true long reading-passage question type. All other additions to the test consist of new *content*, rather than new *question types*. Both multiple-choice and student-produced–response math questions ("grid-ins") now feature algebra II concepts. Same question type, new content. Critical Reading features one fiction passage per test as well as questions on genre, rhetorical devices, and cause and effect. Same question type, different content.

Even the much-feared new Writing section is in a sense old news. The PSAT and the SAT II Writing tests have featured exactly the same multiple-choice question types for years. The essay format and scoring rubric are virtually identical to those of the SAT II Writing test. The College Board had no other choice, given how long the test-development process is.

The other major changes are omissions, not additions: Quantitative Comparisons and Analogies have been dumped from the test.

So, in a nutshell, ETS has simply attached an SAT II Writing test to the old SAT, dropped Analogies and Quantitative Comparisons, added some algebra II content and short reading passages, and ensured that some fiction and fiction-related questions were included. That's it.

A USER'S GUIDE

Reading this book will maximize your score on SAT geometry. We've divided up your study into two sections: **Power Tactics** and **Practice Sets**. The Power Tactics will provide you with important concepts and strategies you'll need to tackle geometry on the SAT. The Practice Sets will give you an opportunity to apply what you learn to SAT questions. To achieve your target score, you'll learn:

- The two question types you'll encounter: multiple-choice and student-produced response, as well as the subtypes: **Treasure Maps**, **False Maps**, and **No Maps**
- What the test-makers are actually trying to test with each geometry question type
- Essential concepts and powerful step methods to maximize your score
- Test-taking strategies that allow you to approach each section with the best possible mindset
- The 10 most common mistakes and how to avoid them

In order to get the most out of this book:

- Make sure to read each section thoroughly and carefully.
- Don't skip the Guided Practice questions.
- Read all explanations to all questions.
- Go to **testprep.sparknotes.com/powertactics** for a free full-length diagnostic **pretest**. This test will help you determine your strengths and weaknesses for geometry and for the entire SAT.
- Go back to our website after you've completed this book to take a **posttest**. This test will tell you how well you've mastered SAT geometry and what topics you still need to review.

THE POWER TACTICS

ANATOMY OF SAT GEOMETRY

Even without reading this book or preparing for the SAT in any way, you'd still get some geometry problems right. However, there's a big difference between:

1. Sweating out a problem, breathing a sigh of relief when you finish it, and timidly moving on.
2. Answering a problem, seeing that the next problem contains a particularly juicy diagram, and licking your chops in expectation of an easy kill.

You don't want to sweat through each question. You want to tear SAT geometry questions apart like a brilliant, vicious, question-slaying gorgon whose blood rises at the very mention of Pythagoras. It isn't as hard as you might think to unlock your inner Geo-Beast. The mistake many students make is taking the SAT cold. That's right, no preparation—not so much as a flip through the information booklet.

A true Geo-Beast studies her prey well before pouncing. By familiarizing yourself with every type of geometry question you can encounter on the SAT, you can approach each geometry question coolly and calmly, knowing in advance what needs to be done in order to answer it correctly. It's about switching from survival mode to attack mode. It's attack mode that will help you score high.

In this section, we provide you with an X-ray of SAT geometry. Later on, we'll review the subtypes of questions and specific strategies for approaching each one. By looking at these questions inside and out, you'll know more about how The College Board tests your skills and how to approach each and every question you'll encounter on the test.

There are two types of math questions on the SAT: multiple-choice and student-produced response.

MULTIPLE CHOICE

Here is a typical multiple-choice question and the terms we'll use to refer to its various parts:

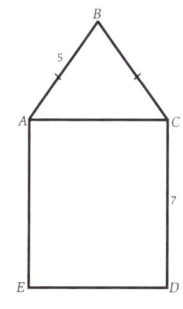

4. If $\angle ABC = 60$ degrees, then what is the area of rectangle $ACDE$?
   (A) $\dfrac{5\sqrt{3}}{2}$
   (B) 5
   (C) 7
   (D) 35
   (E) 49

The sentence containing the question is the **stem**. The lettered options below the stem are called the **answer choices**. Numerical answer choices are always listed in order from smallest to largest or largest to smallest. Only one of these answer choices is correct; the other four answers are called **distractors**, because that's exactly what they're designed to do: *distract* attention from the correct answer. The stem and answer choices grouped together are called an **item**. An entire multiple-choice section, comprised of several items, is called a **set**.

STUDENT-PRODUCED RESPONSE

"Student-produced response" is The College Board's way of saying, "Do it yourself, Bub." Simply put, you, the student, must supply the correct answer without choosing from a group of answer choices. Answering

| ANATOMY OF SAT GEOMETRY |

student-produced responses requires filling in a grid like the one shown below. Therefore we will refer to these questions as **grid-ins**:

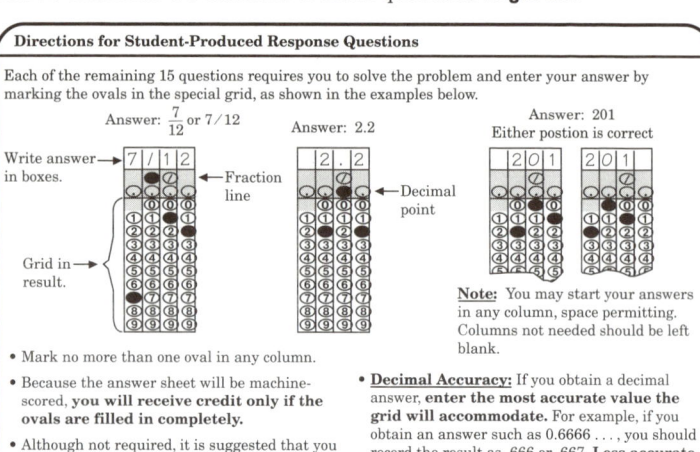

Directions for Student-Produced Response Questions

Each of the remaining 15 questions requires you to solve the problem and enter your answer by marking the oval in the special grid, as shown in the examples below.

Answer: $\frac{7}{12}$ or 7/12   Answer: 2.2   Answer: 201 Either position is correct

Write answer in boxes. ← Fraction line
← Decimal point

Grid in result.

Note: You may start your answers in any column, space permitting. Columns not needed should be left blank.

- Mark no more than one oval in any column.
- Because the answer sheet will be machine-scored, **you will receive credit only if the ovals are filled in completely.**
- Although not required, it is suggested that you write your answer in the boxes at the top of the columns to help you fill in the ovals accurately.
- Some problems may have more than one correct answer. In such cases, grid only one answer.
- No question has a negative answer.
- **Mixed numbers** such as $2\frac{1}{2}$ must be gridded as 2.5 or 5/2. If 2|1|/|2 is gridded, it will be interpreted as $\frac{21}{2}$, not $2\frac{1}{2}$.)

- **Decimal Accuracy:** If you obtain a decimal answer, **enter the most accurate value the grid will accommodate.** For example, if you obtain an answer such as 0.6666..., you should record the result as .666 or .667. **Less accurate values such as .66 or .67 are not acceptable.** Acceptable ways to grid $\frac{2}{3}$ = .666...

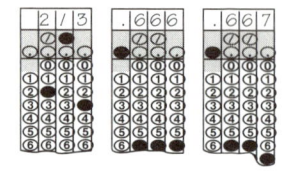

An example of a grid-in might be:

8. Let $f(x)$ be defined as $f(x) = -2x + 4$. Write down the x-coordinate of a point that can be found below this function.

The grid is fairly self-explanatory. If you work out an item and the answer is **2**, you write 2 in the space and then fill in the "2" oval underneath. There are also decimal points and fraction bars in case your answer is not a whole number. We will refer to an individual grid-in as an **item**. A complete grid-in section, comprised of items, is called a **set**.

There are three peculiar things about grid-ins:

1. There may be more than one correct answer to each item. You're probably stuck in the "only one correct choice" mindset brought on by excessive multiple-choice preparation. But don't let this paralyze you: if you get more than one correct answer, pick one, grid it in, and move on to the next item.

| ANATOMY OF SAT GEOMETRY |

2. **Answers can never be negative numbers.** Although there is more than one possible answer, there is actually a limit to what you can grid in. There is no way to denote negative numbers on a grid-in. Why? Who knows, and who cares for that matter? The fact is that all grid-ins must be positive (or zero, which is neither negative nor positive). So if you come up with more than one correct answer, be sure to choose one that is a positive number. If all your answers are negative, you have made a mistake in working out the item.

3. **Improper fractions must be simplified or converted to a decimal answer.** Let's say you come up with $1\frac{1}{2}$ as the answer to an item. If you grid the answer in as $1\frac{1}{2}$, the computer that scans your answer sheet will read your answer as $\frac{11}{2}$. To avoid getting this item wrong, convert the improper fraction into the plain old fraction $\frac{3}{2}$ or the decimal 1.5.

KEY FORMULAS

Math sets on the SAT provide you with key geometric formulas in a reference area that looks like this:

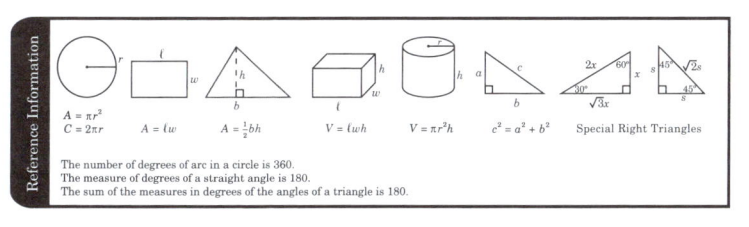

Reference Information

$A = \pi r^2$
$C = 2\pi r$
$A = \ell w$
$A = \frac{1}{2}bh$
$V = \ell w h$
$V = \pi r^2 h$
$c^2 = a^2 + b^2$
Special Right Triangles

The number of degrees of arc in a circle is 360.
The measure of degrees of a straight angle is 180.
The sum of the measures in degrees of the angles of a triangle is 180.

The reference area always appears at the beginning of the set, below the instructions.

WHAT THE SAT COVERS

The geometry items on the new SAT Math section test the following broad concepts:

- Points, lines, and angles
- Triangles
- Polygons
- Circles
- Solids
- Coordinate Grids

| ANATOMY OF SAT GEOMETRY |

- Trigonometry
- Graphs, functions, and transformations

Don't worry if you don't recognize some of these concepts. That's why you're reading this book! Your job is to master two separate areas of expertise:

1. The concepts listed above, which we'll cover in the Essential Concepts section
2. The step methods and strategies, which we'll cover in the Essential Strategies section

ORDER OF DIFFICULTY

The number of an item clues you in to whether it's an easy (low number) or hard (high number) item. Sample items in this book have numbers that approximate where the item would appear on a real SAT Math section. Make sure to note the number of the item before tackling it. We cover order of difficulty in more detail in the Test-Taking Strategies section.

ESSENTIAL CONCEPTS

If SAT geometry were a radio station, then this section would be the top-forty countdown. It's not going to include every fact, just the ones that get the most airplay on the test. When we describe a concept, we cover only the geometry you need to solve SAT items.

You want to have this knowledge down cold. The better you know it, the easier your life will be. Once you're familiar with these concepts, we'll practice applying your knowledge to real SAT items following our specific step methods.

POINTS, LINES, AND ANGLES

Imagine staring through a microscope at a red blood cell for an hour. It would be pretty boring. All alone, blood cells are pretty ho-hum. But you can't really talk about the human body without mentioning blood, can you?

Points

Points are the red blood cells of geometry. By themselves, they're pretty dull, but you can't really talk about larger geometric figures without using points. Lines, triangles, polygons, and angles are all described using the points that comprise them.

For the record, a point technically has no width and length. It's just a place marker. In the figure below, point F is outside the square denoted by the points *MNOP*. Point G is inside the square.

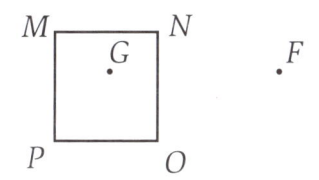

| **ESSENTIAL CONCEPTS** |

Lines

The next step up on the geometric food chain is the **line**. A line is an infinite set of points assembled in a straight formation. Lines have no width, but they do have length. Lines go on infinitely, which we indicate visually by placing a double-arrowed hat over two points that define the line, like this: $\overleftrightarrow{AB}$.

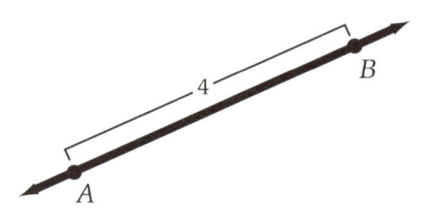

Lines can be **parallel**, meaning they never meet. How sad. Parallel lines are denoted like this: $\overleftrightarrow{AB} \parallel \overleftrightarrow{RS}$. If a line crosses another line at a 90-degree right angle, the two lines are **perpendicular**. In symbol speak: $\overleftrightarrow{TU} \perp \overleftrightarrow{RS}$.

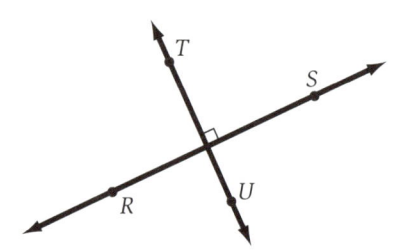

When two parallel lines are cut by a third straight line, that line, known as a **transversal**, will intersect with each of the parallel lines.

Line $\overleftrightarrow{AB}$ goes on forever, but a **line segment** (denoted AB or $\overline{AB}$) does not. Line segments don't get to wear the cool, double-arrowed hat, but they do have a finite length: line segment $AB = 4$. Most geometric figures are made up of line segments.

| ESSENTIAL CONCEPTS |

Rays

If a line and a line segment had a child, it would be a **ray**.

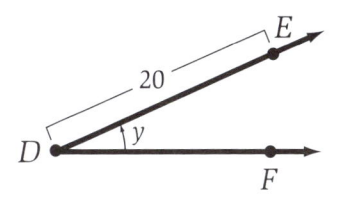

Rays extend infinitely in one direction only, and they wear the not-as-cool, single-arrowed hats. $\overrightarrow{DE}$ and $\overrightarrow{DF}$ are both rays. Notice that you have to put the arrow point over the letter that is heading off into infinity. Point D—called the **endpoint** of the ray—isn't going anywhere. It doesn't get an arrow.

Angles

An **angle** consists of two lines, rays, or line segments sharing a common endpoint. This is very useful to know, since determining the value of angles is a game you will play over and over again on the SAT.

Angles are denoted by the symbol $\angle$. There are many different ways to name an angle, and the SAT will use every one of them. You can use points, but the center letter has to be the **vertex** (the place where the rays start) of the angle. In the previous example, $\angle EDF = \angle FDE$: they are both the same thing.

Angles are measured in **degrees**, which have nothing to do with the temperature. Geometric degrees are indicated by this little guy: °.

| ESSENTIAL CONCEPTS |

Types of Angles

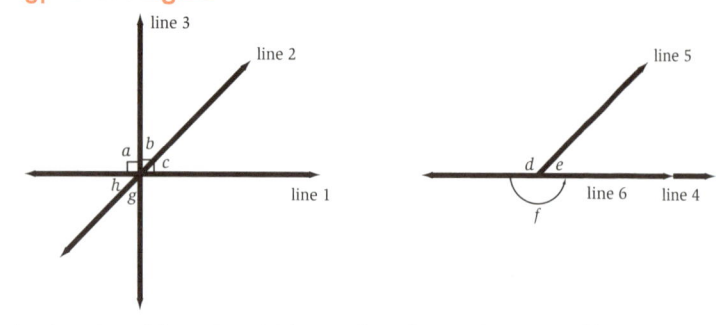

Angle *a* is a **right angle,** which equals 90°. You'll see the ⌐ sign to show that it's a right angle. (This sign is also on the perpendicular lines on page 18.) Combined, angles *b* and *c* also form a right angle. Angles that sum up to 90° are called **complementary** angles.

Lines 2 and 3 cross to make angles *b* and *g*. These are **vertical** angles, and vertical angles are always equal. Angles *c* and *h* are another pair of vertical angles. Shifting over to the other diagram, angle *e* is an **acute** angle (less than 90°) while angle *d* is **obtuse** (more than 90°). There are 180° in a line, so angle *f* = 180°. Angles *d* and *e* add up to make a line, which earns them the moniker **supplementary** angles. Supplementary angles always add up to 180°.

An angle with a measure of 0° is called a **zero angle**. Lines 4 and 6 overlap each other, so they create a 0° angle.

You'll use these basic facts to bop around multiple figures, determining angle values wherever you can. These geometry terms are also used extensively to accurately describe geometric figures.

TRIANGLES

Triangles, triangles, triangles. Get used to the word, because triangles are the most common figure you will see and use on SAT geometry items. Triangles are closed figures containing three angles and three sides.

There are three important rules to know about triangles on the SAT:

1. The sum of the three angles in a triangle will always equal 180°. You'll use this fact so many times on the SAT, you may as well tattoo it to your inner eyelid right now.

| ESSENTIAL CONCEPTS |

2. The longest side of a triangle is always opposite the largest angle; the second-longest side is always opposite the second-longest angle; and the shortest side is always opposite the shortest angle.
3. No side of a triangle can be as large or larger than the sum of the other two sides.

If you know that a triangle has sides of length 4 and 6, you know the third side is smaller than 10 (since 6 + 4 = 10) and bigger than 2 (since 6 − 4 = 2).

Here are three different types of triangles:

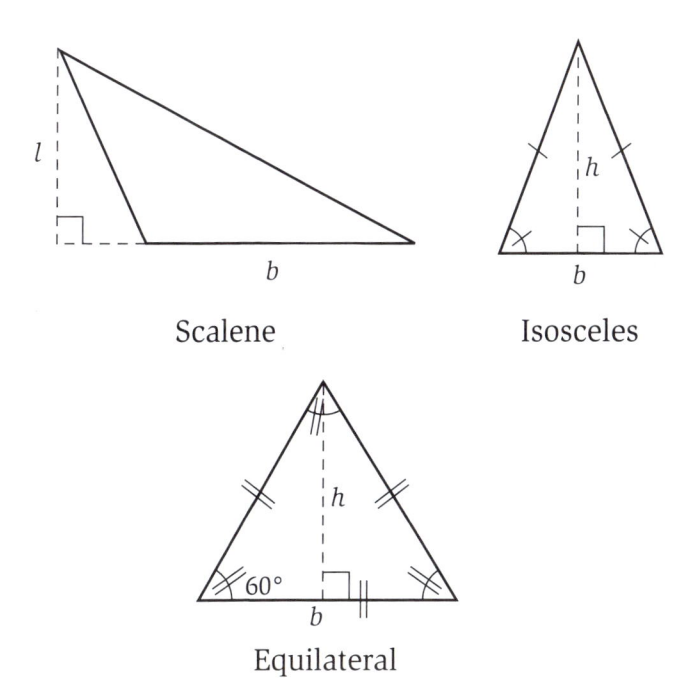

Scalene

Isosceles

Equilateral

Scalene triangles have no equal sides and therefore no equal angles. There's nothing very fiendish or clever about them, so scalene triangles don't get much play on the SAT.

Isosceles triangles have two equal sides, which means they also have two equal angles (the little marks on two of the sides and angles show that they are **congruent**, or equal). Because two of the angles of an isosceles triangle are equal and because all triangles contain exactly 180°, if you know the value of one of the two equal angles, you can figure out the value of all the angles. Isosceles triangles appear in many items on the SAT.

| ESSENTIAL CONCEPTS |

Equilateral triangles are a model of uniformity. They have three equal sides and therefore three equal angles, and the interior angles are always 60° (60 × 3 = 180).

The formula for the area of a triangle is $A = \frac{1}{2}(\text{base})(\text{height})$. You can use any side for the base, but once you pick a base, the height has to be a line perpendicular to the base that extends up to the third point of the triangle. On the scalene triangle, you can see that the height is uneven inside the triangle.

Right Triangles and Pythagoras

A **right triangle** is a triangle with one 90° interior angle (called a **right angle**). Because the angles of a triangle must total 180°, the nonright angles in a right triangle must add up to 90° (180 − 90 = 90). This means that the right angle is always the largest angle in a right triangle. This also means the side opposite the right angle, called the **hypotenuse**, is the longest side in a right triangle. The Greek mathematician Pythagoras figured out that if two sides of a right triangle were known, the value of the third side could also be determined. Of course you had to use his equation to do it:

$$a^2 + b^2 = c^2\text{, where } c \text{ is the hypotenuse.}$$

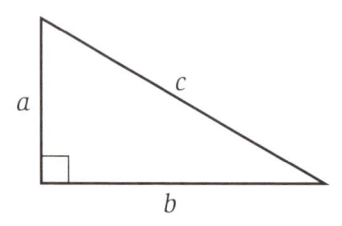

$$a^2 + b^2 = c^2$$

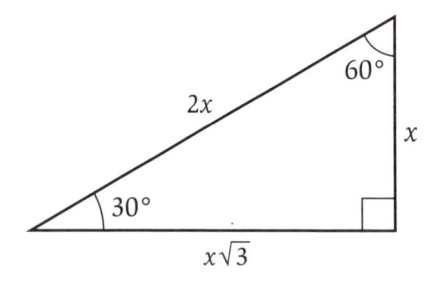

30-60-90 right triangle

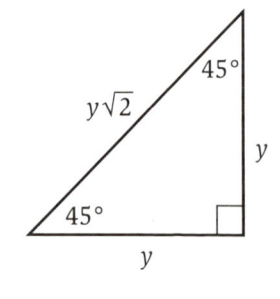

45-45-90 right triangle

| ESSENTIAL CONCEPTS |

Looking at the first triangle, if $a = 3$ and $b = 4$, we can determine what c equals.

$$a^2 + b^2 = c^2$$
$$3^2 + 4^2 = c^2$$
$$9 + 16 = c^2$$
$$25 = c^2$$
$$5 = c$$

This right triangle is called a 3:4:5 right triangle after the length of the sides. Since all three sides are nice round numbers, the SAT likes this right triangle. It also likes the two special-case right triangles shown below it, the 30-60-90 right triangle and the 45-45-90 right triangle.

The 30-60-90 triangle is actually half of an equilateral triangle. If you imagine an equilateral triangle and then cut it down the middle, you'll end up with a 30-60-90. As the second figure shows, the ratio between the three sides is always the same. The side opposite the 90° angle is always twice as long as the side opposite the 30° angle. The side opposite to the 60° angle is always $\sqrt{3}$ times as long as the side opposite the 30° angle.

Learn these two special-case right triangles by heart. Tattoo them to your inner eyelid if you have any extra space.

POLYGONS

A **polygon** is a fancy math name for a closed figure with three or more sides. A triangle is a polygon, but two lines crossing each other are not. A square is a polygon, but a figure in the shape of the letter *C* isn't, because it's open on the right side. Many of the polygons you'll encounter will be **regular** polygons. These Joe Schmo polygons have sides of equal lengths and congruent angles.

Whenever you start talking shapes, you need to talk about area and perimeter. Many items will ask something like, "What is the area of figure so-and-so?" or "What is the perimeter of polygon blah-de-blah?" The **area** of a polygon is all the space inside the figure, while the **perimeter** is the distance around the outer edge of the figure.

Aside from triangles, most of the polygons on the SAT will be **quadrilaterals**. A quadrilateral is any polygon with four sides. Here's a nifty chart about four-sided polygons, area, and perimeter. Many of these formulas

ESSENTIAL CONCEPTS

can be found in the reference portion of the Math test, but it's better to know them from memory.

Quadrilaterals

Type of Polygon	Looks Like	Definition	Area Formula	Perimeter Formula
rectangle	(rectangle with width w and length l)	Opposite sides are parallel and of equal length; interior angles are all right angles; diagonals are equal in length; one diagonal creates two 30-60-90 right triangles	$A = lw$	$P = 2(l + w)$
square	(square with side s)	All sides of equal length; opposite sides parallel; all four interior angles are right angles; diagonals are equal in length; diagonals bisect each other at right angles; one diagonal creates two 45-45-90 right triangles; two diagonals create four 45-45-90 right triangles	$A = s^2$	$P = 4s$
parallelogram	(parallelogram with base b, height h, side c)	Opposite sides are parallel and of equal length; opposite angles are equal; adjacent angles are supplementary; interior angles can be any value; diagonals bisect each other; one diagonal creates two congruent triangles; two diagonals create two pairs of congruent triangles	$A = bh$	$P = 2(b + c)$
trapezoid	(trapezoid with bases b_1, b_2 and height h)	One set of opposite sides are parallel; these two sides, called the bases, are often of unequal length	$A = \dfrac{b_1 + b_2}{2} h$	$P =$ add 'em up!
rhombus	(rhombus with side s)	Four sides of equal length; diagonals bisect each other to form perpendicular lines; diagonals bisect the vertex angles	$A = \dfrac{1}{2} xy$ (where x and y are the lengths of the two diagonals)	$P = 4s$

| ESSENTIAL CONCEPTS |

Hidden Figures Inside Polygons
The first three polygons—square, rectangle, parallelogram—are the big hitters, while the others may only make a guest appearance on an item or two. In addition to having nifty formulas for area and perimeter, these polygons also do a good job of hiding other geometric figures. Put a diagonal inside a rectangle, and it makes two right triangles. Put a **diagonal** in a square, and it makes two 45-45-90 right triangles. You can also chop up a parallelogram or a trapezoid to get a rectangle flanked by two triangles. Don't take our word for it: do it yourself and see.

Interior Angles
Another formula for polygons concerns the sum of the interior angles. For an n-sided polygon, the sum of the interior angles is found with the formula $(n - 2)180 =$ sum of the interior angles. If you have a triangle, the sum is: $(n - 2)180 = (3 - 2)180 = (1)180 = 180$, but you knew that already. For a seven-sided **heptagon**, the sum is: $(n - 2)180 = (7 - 2)180 = (5)180 = 900$. You can contrast this formula for interior angles with the fact that the sum of the exterior angles of any polygon is always 360°.

Know your polygons. 'Nuff said.

CIRCLES

After so many cruel, rigid line drawings, it's nice to contemplate the **circle**, nature's roundest creation:

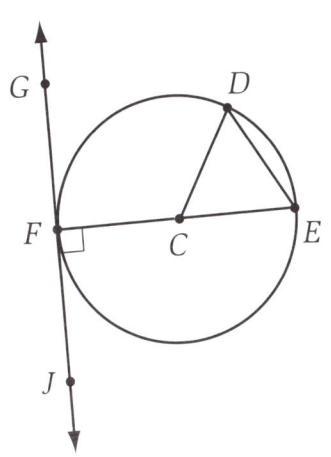

| ESSENTIAL CONCEPTS |

This is a circle with **center** C. All points along the edge are equidistant from C. This means that line segments $\overline{CD} \cong \overline{CE} \cong \overline{CF}$. The equals sign with a squiggle over it is the symbol for **congruence**, the geometrical way of saying that the three line segments are all equal. Any line going from the center of the circle to the edge is called a **radius**. $\overline{CD}$, $\overline{CE}$, and $\overline{CF}$ are all radii. If a line goes from the edge, through the center, and continues on to the other edge, it is called a **diameter**, and its length is equal to two radii. The diameter bisects a circle. $\overline{FE}$ is a diameter.

The **circumference** of a circle is the distance around the outer edge, so circumference is like saying "perimeter of a circle." The circumference and area formulas for a circle are:

$A = \pi r^2$
$C = 2\pi r = \pi d$

Both of these formulas contain the mathematical constant **pi**, which has a value of around 3.14 and looks like this: π, and the variable r for radius (in the circumference formula you can also use the variable d for diameter, which is equal to $2r$). This means that if you're given the circumference of a circle, you can find its area by first solving for r using the circumference formula and then plugging that value of r into the area formula. You can use the area of a circle to find its circumference by doing the same thing: solving for r and then placing that value into the proper formula. Test-makers think this is very clever, so you should expect to do it on the SAT.

Tangent Lines

Properties of **tangent lines** are one of the concepts covered on the new SAT. Look at line GJ. It touches the circle at only one point, F. At the point where a tangent touches a circle, the tangent line is perpendicular to the circle.

Whenever a tangent hits a circle, it creates a right angle with a radius of the circle. Earn some brownie points and think of what the SAT could do with this fact. Draw a dotted line from G to C, and draw another from J to C. What you get is two right triangles.

| ESSENTIAL CONCEPTS |

Arcs, Sectors, and Some Other Concepts That Sound Harder Than They Are

It's slice-and-dice time!

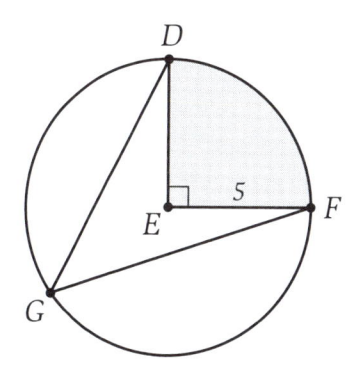

Angle DEF has its vertex at the center of circle E, so it shouldn't blow your mind that this is called a **central angle**. Since ∠DEF is a right angle, the **arc** (portion of the circle) DE sliced out by this angle is equal to 90°. An arc defined by a central angle is always equal to the measure of that angle.

Contrast this to an **inscribed angle**, which is an angle formed by two **chords** (lines from one point on the circle to another that do not pass through the center) with a vertex resting on the edge of the circle. Angle DGF is an inscribed angle formed by chords GD and GF. An inscribed angle is half the size of the arc it creates. In our diagram above, we already know that arc DF is 90°. Therefore, angle DGF must be half of 90°, so ∠DGF = 45°. This little fact makes inscribed angles quite sneaky, so if you do encounter one, it will probably be on a medium or hard item.

Arc lengths and **sectors** are pretty straightforward concepts:

Arc length = portion of the circumference.

Sector = portion of the area.

You don't have to be Stephen Hawking to realize that before you solve for arc length, you have to solve for the circumference. Circle E has a radius of 5, so the circumference is:

$$C = 2\pi r$$
$$C = 2\pi 5$$
$$C = 10\pi$$

| ESSENTIAL CONCEPTS |

Now, arc *DF* takes up 90°, and 90° is *one-fourth* of the total degrees in a 360° circle, so (drumroll) . . . arc length *DF* will be *one-fourth* of the total circumference.

$$\text{Arc Length} = (C)\left(\frac{\text{arc}}{360}\right)$$
$$\text{Arc Legnth} = (10\pi)\left(\frac{90}{360}\right) = 10\pi\left(\frac{1}{4}\right) = \frac{5\pi}{2}$$

Sector is the area equivalent of arc length. Look at the shaded portion of the circle, which takes up one-fourth of the total circle since the central angle is 90°. To find the area of the shaded sector, find the area and multiply it by whatever fraction of the total circle the sector comprises.

$$\text{Sector} = (\text{area})\left(\frac{\text{central angle}}{360}\right)$$
$$\text{Sector} = (\pi r^2)\left(\frac{90}{360}\right)$$
$$\text{Sector} = (\pi 5^2)\left(\frac{1}{4}\right)$$
$$\text{Sector} = (25\pi)\left(\frac{1}{4}\right)$$
$$\text{Sector} = \frac{25\pi}{4}$$

Arc lengths and sectors are not basic geometry concepts, and their weirdness is what makes them tough for many students. But their bark is worse than their bite: once you learn what they mean, finding the correct value simply requires multiplying the total area or circumference by a fraction of the total circle.

SOLIDS

2 + 1 = 3. Simple enough. Take any two-dimensional polygon and add one more dimension to it (**height**), and you have a three-dimensional **solid**.

A **right circular cylinder** is just a circle with height. A **cube** is a square increased upward by the length of one of its sides. A **rectangular solid**—which most humans would call a *box*—is a rectangle with height added to it.

| ESSENTIAL CONCEPTS |

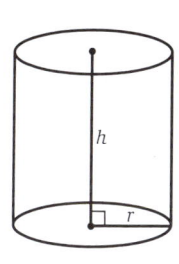

$V = \pi r^2 h$
right circular
cylinder

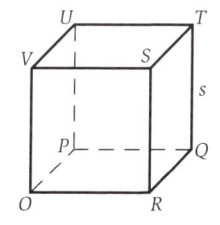

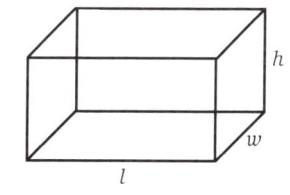

$V = s^3$ or lwh, where $l = w = h$
cube

$V = lwh$
box

There are three main item types associated with three-dimensional figures:

1. Volume items, designed to blow the minds of students used to area and perimeter
2. Hidden shape items, like the well-concealed right triangle *UPR* hiding within the cube
3. Surface area items, which sometimes throw out the term *face*

The volume items shouldn't cause you much trouble. The volume formulas are in the reference portion of the Math test, so check them over if you're unsure.

The sneaky hidden-shape items are a little tougher but not insurmountable by any means. The box and cube are swarming with triangles, sure, but they are also chock-full of right angles. This means all the triangles are right triangles, so you can use the good ol' Pythagorean Theorem.

If the box in our diagram were somehow alive, the **surface area** would be its skin. The surface area of a rectangular solid equals the sum of the area of the six rectangles that make up its sides. For a cube, the surface area would be the sum of the area of its six sides. These sides are sometimes referred to as **faces**. Written in geometry language:

| **ESSENTIAL CONCEPTS** |

Surface area of a rectangular solid =
2(length)(width) + 2(length)(height) + 2(height)(width)

Surface area of a cube = 6(side)^2

There are formulas for figuring out the surface area of other solids, but the SAT will only ask you to calculate the surface area of a rectangular solid or a cube.

COORDINATE GRIDS

Every math concept covered so far can be placed on a coordinate grid. Standard two-dimensional coordinate grids are set up with a horizontal **x-axis** and a vertical **y-axis**. The place where these lines meet is the **origin**. At the origin, $x = 0$ and $y = 0$. Values of x left of the y-axis are negative, while those to the right are positive. Values of y above the x-axis are positive, while those below it are negative.

Any point can be expressed on the grid using the formula (x,y), so the origin is always (0, 0).

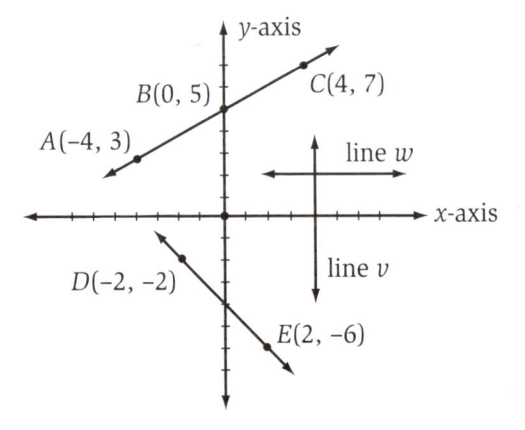

Point D is at (−2, −2), meaning it's two spaces to the left and then two spaces down. Point C is at (4, 7), so you go four spaces right and then seven spaces up.

Lines appear on grids quite a bit. On a graph, an equation of a line usually takes the form $y = mx + b$. The variables x and y stand for the (x, y) of any point on the line, while m is the **slope** of the line and b is the **y-intercept**.

| ESSENTIAL CONCEPTS |

Slope

The slope of a line, denoted by the variable m, is the change in y-values divided by the change in the x-values of the line. The slope can be determined from any two points on a line. Take points C (4, 7) and B (0, 5), which are both on line AC.

$$\text{slope} = m = \frac{y_1 - y_2}{x_1 - x_2}$$

$$m = \frac{7-5}{4-0} = \frac{2}{4} = \frac{1}{2}$$

You can use different points on the same line, like A and B, but the slope will still be the same:

$$m = \frac{y_1 - y_2}{x_1 - x_2} = \frac{3-5}{-4-0} = \frac{-2}{-4} = \frac{1}{2}$$

Viewed from left to right, slopes are positive if the line is moving upward and negative if the slope heads down. Without knowing the exact value, you know the slope of line DE is negative. Some people like to remember slope as "rise over run." Starting at point A, if you "rise" one space and then "run" two spaces over, you find yourself at L, another point on the line with a slope of $\frac{1 (\text{rise})}{2 (\text{run})}$.

The slope of line w is zero, while the slope of line v is undefined. Plug some points into the slope formula to see for yourself. Or just take our word for it. We never lie about slope.

y-Intercept

The b in $y = mx + b$ is called the y-intercept, and it's the place the line crosses the y-axis. When a line crosses the y-axis, the value for x is zero. Watch what happens when x equals zero:

$$y = mx + b$$
$$y = m(0) + b$$
$$y = b$$

Understanding all the parts of the linear equation $y = mx + b$ (especially slope) will help you answer many coordinate grid items. But since anything can be placed on a graph, these items will not be limited to finding

| **ESSENTIAL CONCEPTS** |

the slope. Suppose you were given the diagram below and then told *H* is a point on the line *DE*:

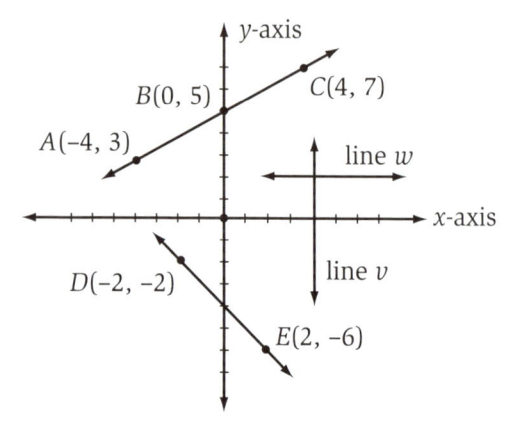

If *H* is at (1, −5), how long would line segment *DH* be?

First, place *H* on the grid at (1, −5). You can see a right triangle with *DH* as its hypotenuse. It's even a 45-45-90 right triangle:

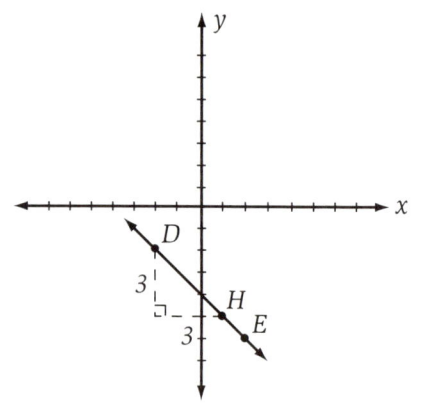

The answer, by the way, is $3\sqrt{2}$.

Midpoint

To find the **midpoint** of line segment *DH*, use the midpoint formula:

$$\text{Midpoint} = \left(\frac{x_1 - x_2}{2}, \frac{y_1 - y_2}{2}\right)$$

There's nothing wildly exciting about this formula. You just take the *x* and *y* values for the two endpoints and find the average for each one.

| ESSENTIAL CONCEPTS |

TRIGONOMETRY

One of the topics making its debut on the new SAT is basic **trigonometry**. However, The College Board tells us that "these questions can be answered using trigonometric methods, but may also be answered using other methods."

Sheesh. So you could spend tons of time learning a difficult subject, or you could just use your knowledge of triangles to answer these items. Look at this figure:

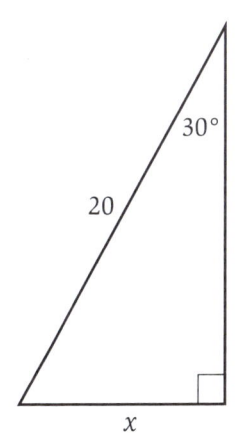

You could solve for x using the trigonometric equation $\sin 30° = \frac{x}{20}$, or you could use your knowledge of 30-60-90 right triangles to determine that the leg opposite the 30° is half the hypotenuse. Either way your answer would be 10. The important thing to remember here is that any trig item on the SAT can be solved using basic geometry.

Life's too short. If you don't already know trig, forget learning it for the SAT. Move on!

GRAPHS, FUNCTIONS, AND TRANSFORMATIONS

This topic seems impossibly complicated, but you can get through it if you do one simple thing: ignore all the fancy terms and just act like a robot.

Here is an example of a **function**: $f(x) = \frac{1}{2}x + 5$. For every value of x you put in the right side of the equation, the function spits out one (and only one) corresponding value. Pretty robotic. Let's jam $x = 0$ into the function and see what comes out.

ESSENTIAL CONCEPTS

$$f(x) = \frac{1}{2}x + 5$$
$$f(x) = \frac{1}{2}(0) + 5$$
$$f(x) = 5$$

So if x is zero, the corresponding value of $f(x)$ is 5.

Now, on a graph, $f(x)$ acts as the y-value, so you've just discovered the point (0, 5). Seem familiar? It should, since that's point B, your y-intercept on page 29. The function is nothing more than the equation for line AC in the form $y = mx + b$, only $f(x)$ is masquerading as y.

Think about it. If you understand linear equations in the form $y = mx + b$, then you understand how a linear function appears on a graph. Granted, not every function is going to be a straight line: you can get some funky functions if you start squaring terms. In general, though, whatever function you're given, put in values for x and see what values for y are spit out. Then graph them.

Transformations

Transformations can be tricky. Again, acting robotic can make them easier. You'll be given an initial function, and then this function will be "transformed" in some manner. Here's an example using a linear function:

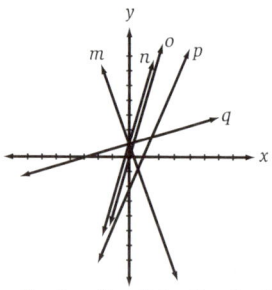

7. Which line shows the function $f(x) = 3x + 1$ after $f(x)$ has been transformed to $f(x - 1)$?

(A) line m
(B) line n
(C) line o
(D) line p
(E) line q

| ESSENTIAL CONCEPTS |

All right, robot, here are your instructions:

1. Take $(x-1)$ and place it into the original function.
2. Solve this new equation.
3. Using the new slope and y-intercept of this "transformed function," locate the new line.

You can replace $f(x)$ with y anytime you want.

$$f(x) = 3x + 1$$
$$f(x) = 3(x-1) + 1$$
$$f(x) = 3x - 3 + 1$$
$$y = 3x - 2$$

The y-intercept (b in $y = mx + b$) is -2, so the new, transformed line has a point at $(0, -2)$. Only line p crosses that point, so **D** is the answer.

That's a transformation. Good work. Power down, robot.

Parabolas

Imagine if the world's strongest person took a line, grabbed it in both hands, and then somehow bent it. The resulting shape would be a **parabola**, a U-shaped curve that can open either upward or downward on a coordinate graph.

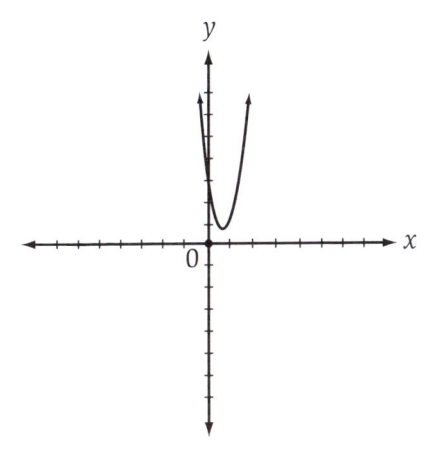

Parabolas occur when the x value in a function is squared. A common equation for a parabola is: $y = ax^2 + bx + c$. The equation for our parabola is $y = 4x^2 - 6x + 3$.

You can see this is similar to the equation for a line, $y = mx + b$, only we've added a squared term to the front of it.

| **ESSENTIAL CONCEPTS** |

Parabolas have many characteristics. The two key ideas you have to learn for the SAT are how to:

1. Determine whether the parabola opens upward or downward
2. Locate the **vertex**—the bottom point of the *U*—of the parabola

To determine whether the parabola opens upward or downward, look at the *a* value (the number in front of the squared term). The parabola shown in the diagram conforms to the equation $y = 4x^2 - 6x + 3$. The *a* value, 4, is positive, which is why the parabola is heading upward.

The vertex is a bit trickier. This point is found by the very complicated formula:

$$\text{Vertex Point} = \left(\frac{-b}{2a}, c - \frac{b^2}{4a}\right)$$

It would be nice if this formula was at the front of every Math section, but no such luck. For the parabola $y = 4x^2 + 6x + 3$, the vertex ends up at $\left(\frac{3}{4}, \frac{3}{4}\right)$. If you don't believe us, block out about five minutes of time and work it out yourself.

Thankfully, that's the last geometry concept we're going to cover. Now that your mind is crammed with the right facts, it's time to show you the best way to put this knowledge to use.

ESSENTIAL STRATEGIES

To capture the essence of SAT geometry, look at the following figure, which resembles a child's drawing of a house:

Here's how most SAT geometry works:

- If you get information about a triangle, you'll be asked something about a rectangle.
- If you get information about a rectangle, you'll be asked something about a triangle.
- When a diagram is supplied, you'll have to make a leap from information about the figure to information supplied in the stem.

This is SAT geometry at its most simplistic, so let's dress it up to make it look a little more realistic. When a diagram is supplied, information is given about the diagram in two places: on the diagram itself and then in a crucial morsel placed within the stem. An actual SAT geometry item would look like:

| ESSENTIAL STRATEGIES |

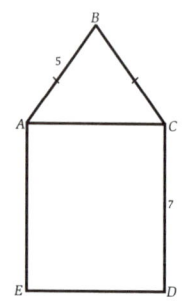

4. If $\angle ABC = 60°$, then what is the area of rectangle $ACDE$?

(A) $\dfrac{5\sqrt{3}}{2}$
(B) 5
(C) 7
(D) 35
(E) 49

In this example, you are given information about the triangle but asked about the rectangle. There's information about the triangle in the diagram, but you can bet it's not enough to answer the item. Another fact is given in the stem ($\angle ABC = 60°$), but it's not enough to answer the item alone. It is the information about the triangle in the diagram *combined with* the information in the stem that will provide you with what you need to find the area of the rectangle.

TYPES OF GEOMETRY ITEMS

On the SAT, geometry items are one of three basic types:

- Treasure Maps
- False Maps
- No Maps

Treasure Maps
The majority of SAT geometry items you encounter will feature a drawn-to-scale diagram. Think of these as Treasure Map items. The diagram is the "map" that leads you to the correct answer, a treasure for those hoping to boost their SAT score.

The example we used above is a typical Treasure Map item.

| ESSENTIAL STRATEGIES |

False Maps

Beware the dreaded words *Note: Figure not drawn to scale* on an item. Figures not drawn to scale will distort items, which makes it tougher to make the correct mathematical deductions:

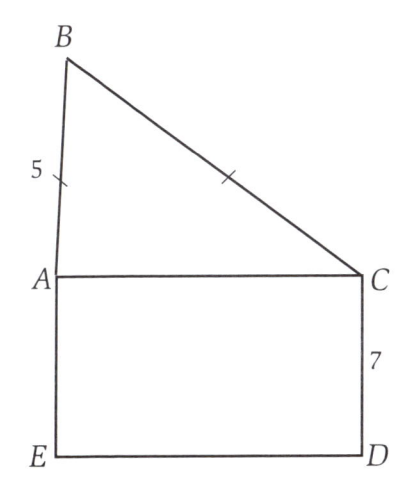

Note: Figure not drawn to scale.

The lines of the triangle no longer look even. This makes it harder to visualize the item, which in turn makes it more difficult to take the correct steps needed to answer it.

No Maps

Look at this item:

7. Points $A, B,$ and C are the vertices of a triangle. Points $A, C, D,$ and E are the vertices of a rectangle. The length of AB is 5 and the length of CD is 7. Sides AB and BC are congruent. If $\angle ABC = 60$ degrees, then what is the area of rectangle $ACDE$?

(A) $\frac{5\sqrt{3}}{2}$
(B) 5
(C) 7
(D) 35
(E) 49

This is the same Treasure Map item shown above, without a diagram. In other words, no map.

| ESSENTIAL STRATEGIES |

TACKLING TREASURE MAPS

Treasure Maps are SAT geometry items with a diagram. To do well on these items, you must learn how to fit incomplete pieces of information together to form a conclusion. That's how SAT geometry works, plain and simple. It's how 90 percent of all geometry items on the test are designed. Understand the process, and you'll have SAT geometry licked.

Here's a four-step method for approaching geometry items with a diagram in them:

Step 1: Combine the information in the diagram with the information in the stem.

Step 2: Determine what information is needed to answer the item.

Step 3: Use basic geometry knowledge to determine values.

Step 4: Give the item what it wants.

Treasure Maps in Slow Motion
Let's look at each step more closely, using the same example item:

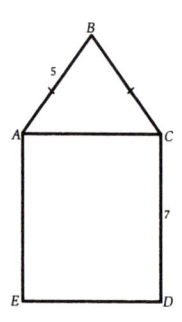

4. If $\angle ABC = 60°$, then what is the area of rectangle $ACDE$?

(A) $\dfrac{5\sqrt{3}}{2}$
(B) 5
(C) 7
(D) 35
(E) 49

| ESSENTIAL STRATEGIES |

Step 1: Combine the information in the diagram with the information in the stem.

List all the facts in a single place. The easiest way to do this is to add the written information to the existing diagram. For our house example, make a notation that shows angle *ABC* is 60°. Every fact in the item is now on the diagram.

Step 2: Determine what information is needed to answer the item.

If you skip this step, the item becomes much harder. The stem asks for the area of the rectangle, but you won't find it by only looking at the diagram. It's still well hidden. What you must remember is that the area of a rectangle is found by multiplying its length by its width ($A = lw$). The length of side *CD* is provided, so now you need the width. You need to determine either the width of *ED* or the width of *AC*.

AC is also the side of the triangle. Alarms should start ringing in your head. Your inner SAT voice should be yelling, "There must be a way to take all the information given about the triangle to find the length of *AC*."

Step 3: Use basic geometry knowledge to determine values.

The "basic geometry knowledge" we're talking about is covered in "Essential Concepts." In our example, use the following line of reasoning, which assumes some core geometry knowledge:

- Those little dash marks in *AB* and *BC* mean that they are equal in length.
- Angles that are opposite sides of equal length are also equal, so angle *BAC* and *BCA* are the same.
- Added together, all three interior angles of a triangle always equal 180°.
- If angle *ABC* = 60°, then angle *BAC* and *BCA* must equal 120° combined.
- Since angle *BAC* and *BCA* are equal, they must each be 60°. Every angle in the triangle is 60°, so every angle is equal, which means that every side is also equal.
- Therefore, *AC* must be 5.

You can see how important it is to brush up on the basics. Don't sweat memorizing every formula, but buckle down and memorize all the facts, because you will need to apply this stuff effortlessly on the day of the test.

Step 4: Give the item what it wants.

| ESSENTIAL STRATEGIES |

Don't do all the legwork only to goof at the last second. Make sure you come up with the answer the item wants, not a part of the answer or subset of the answer. For example, on our sample item, a student might mistakenly pick choice **B** since it's the length of AC. However, the stem is really asking for the area of rectangle ACDE. The area is: $A = lw = (7)(5) = 35$, choice **D**. Don't let the four distractors fool you!

The way you approach these four steps will vary according to the item, of course, but the essential process behind them will remain the same. It may take some time to get comfortable with the four steps, but once you do, you'll start salivating whenever you see an SAT geometry item. You'll know what you need to do *before* you even look at the item.

Guided Practice
Try this one on your own:

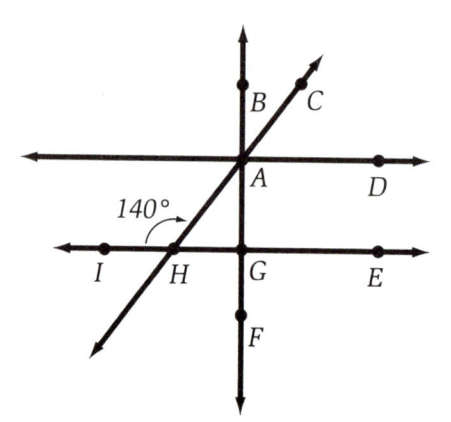

5. If $\angle BAC = 48°$, what is the value of $\angle HGA$?

(A) 40
(B) 48
(C) 90
(D) 92
(E) 140

Step 1: Combine the information in the diagram with the information in the stem.

Remember, you need all the facts together in one place. This will go a long way toward helping you work through the item.

| ESSENTIAL STRATEGIES |

Step 2: Determine what information is needed to answer the item.

What is the stem asking you for? What other values do you need to determine first? Is there anything "hidden" in the diagram?

Step 3: Use basic geometry knowledge to determine values.

Think back to the essential concepts you learned in the previous section. What values can you figure out?

Step 4: Give the item what it wants.

Make sure you answer what the item is asking you to determine. Don't get sidetracked by distractors.

Guided Practice Explanation

A bunch of lines get jammed together, but notice that there are no markings saying that any of these lines are either parallel or perpendicular. Let's use our step method to figure this item out.

Step 1: Combine the information in the diagram with the information in the stem.

The diagram tells you that $\angle AHI = 140°$. From the item, you learn that $\angle BAC = 48°$. Add that fact onto the diagram so that all information is in the same place. Now let's see what we can do.

Step 2: Determine what information is needed to answer the item.

At first glance it appears that we don't have enough information to answer this item. Uh-oh, what do we do? Take a closer look at the diagram. Do you notice anything hidden in the jumble of lines? You should see a hidden triangle marked by points A, G, and H. To find the value of one angle in a triangle, you need to know the value of the other two angles. So, to figure out the value of $\angle HGA$, we have to first know the values of $\angle AHG$ and $\angle HAG$.

Step 3: Use basic geometry knowledge to determine values.

Let's go through our line of reasoning:

- We know that two supplementary angles always equal 180°.
- Since you know $\angle AHI = 140°$, you can determine that $\angle AHG = 40°$.

| **ESSENTIAL STRATEGIES** |

- We know that two vertical angles are always equal to each other.
- Since you know that ∠BAC = 48°, ∠HAG must also equal 48°.

Step 4: Give the item what it wants.

Notice that two of the answer choices are 40 and 48. If you're in a hurry, you might see these familiar numbers and choose **A** or **B**, but these answer choices are actually distractors. See the item all the way through, and you'll reach the right answer.

You now know two of three interior angles of the triangle *HAG*. Since the stem asks for the third angle (*HGA*), all you need to do now is a little math:

$$\angle HGA + \angle AHG + \angle HAG = 180$$
$$\angle HGA + 40 + 48 = 180$$
$$\angle HGA = 92$$

There's your answer, choice **D**.

No part of the stem mentioned the word *triangle*, yet your knowledge of triangles led you to the right answer. It's important that you learn to recognize these hidden treasures wherever they appear. The SAT pulls this trick all the time.

| ESSENTIAL STRATEGIES |

Independent Practice
After you complete the following item, turn the page for the explanation.

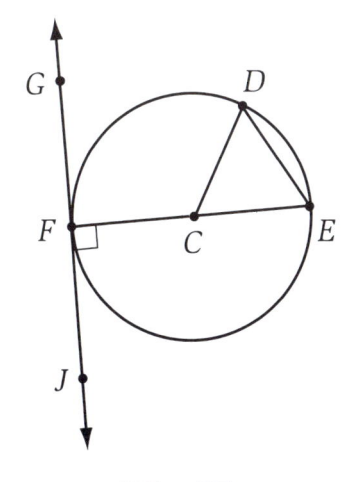

$\overline{DE} \cong \overline{CF}$

7. If the area of circle C is 49π, what is the perimeter of triangle CDE?

(A) 7π
(B) 14π
(C) 21
(D) 42
(E) It cannot be determined from the information given.

| ESSENTIAL STRATEGIES |

Independent Practice Explanation

Step 1: Combine the information in the diagram with the information in the stem.

The stem tells you the area, while the diagram states $\overline{DE} \cong \overline{CF}$. Add the area to your diagram.

Step 2: Determine what information is needed to answer the item.

To find the perimeter of the triangle, figure out the length of each side. None of these values are provided, so you must determine the length of sides *CD*, *CE*, and *DE*.

Step 3: Use basic geometry knowledge to determine values.

Start with the area information first, since that will give you the radius.

- We know that the area of a circle = πr^2.
- If $49\pi = \pi r^2$, then $49 = r^2$.
- So, $7 = r$.
- Looking at the diagram, two of the legs of triangle *CDE* are radii: *CD* and *CE*.
- So, *CD* = 7 and *CE* = 7.
- The diagram tells you that $\overline{DE} \cong \overline{CF}$.
- *CF* is also a radius, so it is equal to 7.
- That means *DE* = 7.

Step 4: Give the item what it wants.

You now have all the values necessary to solve this item. The perimeter of a triangle is the value of its three sides added together. All three sides of triangle *CDE* are length 7—an equilateral triangle, who'd of thunk it? Now for a little addition: 7 + 7 + 7 = 21. The answer is **C**. The distractors in this item aren't too tricky. If you were rushing, you may have thought **A**, 7π, was the right answer. But hopefully the π would have given the distractor away.

    This item is typical of the kind of mathematical tap dancing required on the SAT, so get used to it.

| ESSENTIAL STRATEGIES |

TACKLING FALSE MAPS

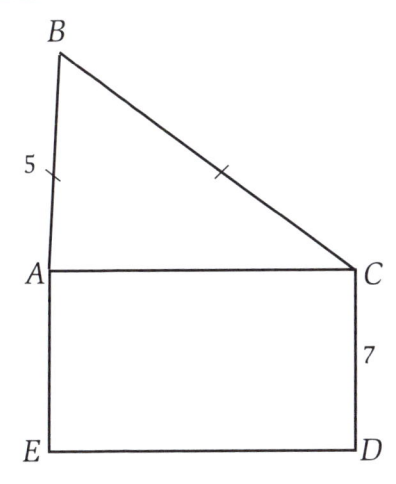

Figure: a house-shaped diagram with triangle roof ABС (B at top, A at lower-left of roof, C at lower-right of roof) atop rectangle ACDE. AB is labeled 5, with slash marks on AB and BC indicating equal length. CD is labeled 7.

Note: Figure not drawn to scale.

Our poor house! Even though the slash marks are still in the two lines making up the roof, they don't look even.

To combat the False Map, change the first step of our approach:

Step 1: Redraw the figure as accurately as possible.

On your new figure, combine all the information from the diagram and the stem into the single picture. You need to stop looking at a figure that is deliberately false, since this false image can impede your ability to answer the item.

On the distorted diagram, the length of *CD* (7) is shorter than *AB* (5). Your new diagram should correct that, and it should also make *AB* and *BC* equal in length.

After redrawing the picture, proceed with the item using the usual Treasure Map steps 2, 3, and 4.

| ESSENTIAL STRATEGIES |

TACKLING NO MAPS

Here is an example of an item with no map:

7. Points A, B, and C are the vertices of a triangle. Points $A, C, D,$ and E are the vertices of a rectangle. The length of AB is 5 and the length of CD is 7. Sides AB and BC are congruent. If $\angle ABC = 60°$, then what is the area of rectangle $ACDE$?

    (A) $\frac{5\sqrt{3}}{2}$
    (B) 5
    (C) 7
    (D) 35
    (E) 49

You might think this item would be simpler to solve than a Treasure Map item since all the information is provided within the stem. But the lack of a corresponding visual makes it tougher to grasp, especially for ultratidy types who don't want to muss up their SAT test booklets with pictures.

If you see a No Map item like the one above, don't try to "visualize" it in your mind and then solve it there. Instead, draw the figure yourself and follow these steps:

Step 1: Draw a diagram.

Be as accurate as possible with your sketch, but don't spend a lifetime getting the shading just right. Having a diagram in front of you will free up precious room in your brain for determining values for lines and angles.

After drawing yourself a picture, you can proceed with the item using the usual Treasure Map steps 2, 3, and 4.

WHAT IF YOU ARE STUMPED?

There will be an item or two where your knowledge of geometry fails you. Don't sweat it. It happens to everyone. There are so many rules, one of them is bound to slip through the mental cracks occasionally. When this occurs, take solace in the directions for the Math section, which state:

> Figures that accompany problems in this test are intended to provide information useful in solving the problems.

| ESSENTIAL STRATEGIES |

On Treasure Map items, this means you can use your eyes to do some Process of Elimination (shortened to "POE" for those of us on-the-go). If the figure is not drawn to scale (i.e., False Map), or there is no figure (i.e., No Map), this method won't work.

Eyeballing

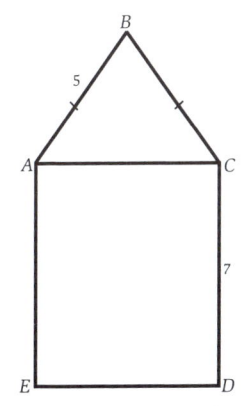

4. If $\angle ABC = 60°$, then what is the area of rectangle $ACDE$?

(A) $\frac{5\sqrt{3}}{2}$
(B) 5
(C) 7
(D) 35
(E) 49

Cast your eyes back to this quaint house item again. To find the area of the rectangle, you need length AC. You already have one length, $CD = 7$. Look at AC and ask, "Does AC look shorter, longer, or about the same as CD?" Your eyes should tell you that it looks a bit smaller. So let's say $AC = 6$, since 6 is a little smaller than 7.

If $AC = 6$, then the area of rectangle $ACDE$ is: $A = lw = (6)(7) = 42$. This isn't the actual answer, but it's a good approximation of it. Armed with this guess, you can see that answer choices **A**, **B**, and **C** are all way too small to be the right answer. Recall that numerical answers are always listed in order from smallest to largest or vice versa, so you don't have to figure out $\frac{5\sqrt{3}}{2}$, choice **A**, to know that it's smaller than 5 and therefore incorrect.

| **ESSENTIAL STRATEGIES** |

Using your eyes helps you bypass traps completely. The answer has to be either **D** or **E**. If it were **E**, then the width and length would both be 7. Your eyes tell you that AC is less than CD, so **E** is not the best guess. Pick **D**, and you'll get the right answer. It won't work this smoothly every time you try eyeballing an item, but you'll often be able to brush off a few choices and then take a guess.

The house example dealt with lengths, but you can also get some POE distance out of eyeballing angles:

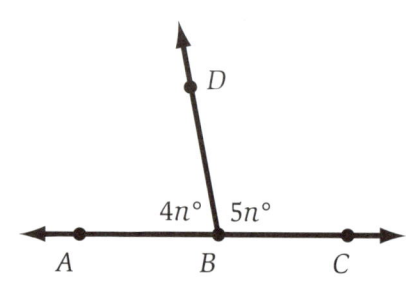

2. If $\overleftrightarrow{AC}$ is a line, which of the following is the measurement of ∠DBC?

    (A)   20°
    (B)   80°
    (C)   100°
    (D)   110°
    (E)   160°

Just eyeball ∠DBC and ask yourself, "Is this angle greater than or less than 90°?" Everyone has a good idea what a 90° angle looks like. ∠DBC is a little more than 90°. Not much, but a bit more. Scanning the answer choices, you can eliminate choices **A** and **E** quickly, since both are way off base. Choice **B**, 80°, is closer, but it's less than 90°, so it's unlikely to be the correct answer. That leaves us with **C** and **D**. There's no way to use your eyes to make a 10° call one way or the other, but for a fifty-fifty chance you can guess, which is well ahead of the wrong-answer penalty. The answer, by the way, is **C**:

TREE STORES, INC.

Store# (507) 289-0708
1201 South Broad
Suite 240
Rochester, MN 55904

DESCRIPTION	QTY	PRICE	TOTAL
BATHROOM CLEANER	1	1.00	1.00T
SCRUB BRUSH	1	1.00	1.00T
SPINACH	1	1.00	1.00N
BISCUIT	1	1.00	1.00N

```
           Total          $4.00
           TRADE EXEM     $0.00
           TAX             0.16
                          $4.16
                         $20.00
```

Thank you for shopping at Dollar Tree
Where Everything's $1.00
Shop Online at Dollartree.com

****
* We value your opinion!   *
* Please provide your feedback!   *
* www.dollartreefeedback.com   *
* Receive chances to win $1,000 daily plus *
* instant prizes valued at $1,500 weekly *
* or by calling 1-877-968-2540.   *
* For complete rules, eligibility and sweepstakes *
* period and previous winners please visit *
* www.dollartreefeedback.com   *
* No purchase/survey required to enter. *
* Sweepstakes sponsored by Empathica Inc. *
* and its multiple international clients. *
* Survey Code: 4555 6869 4432 0003   *
*   *
* We will gladly exchange any unopened item *
* with original receipt. We do not offer refunds. *
****

0459 0325 11 041 46917   4/05/16 16:45
Sales Associate: Jennifer

$$y = \frac{4}{5}(x-5)^2 + 2$$
$$y = \frac{4}{5}x^2 - 10x + 25 + 2$$
$$y = 4x^2 - 50x + 27$$

| ESSENTIAL STRATEGIES |

$$4n + 5n = 180$$
$$9n = 180$$
$$n = 20$$
$$\angle DBC = 5n$$
$$\angle DBC = 100$$

You visually estimate any geometry item that's drawn to scale. Even if you understand the "math" needed, using both approaches to attack an item greatly increases your chance of getting it right. If you don't understand the math, then using your eyes can often give you a good shot at a tough item.

Plugging In for x

If you've tried the step method or eyeballing and neither approach works for you, try plugging in values for x in an equation to see where that gets you. Look at this item:

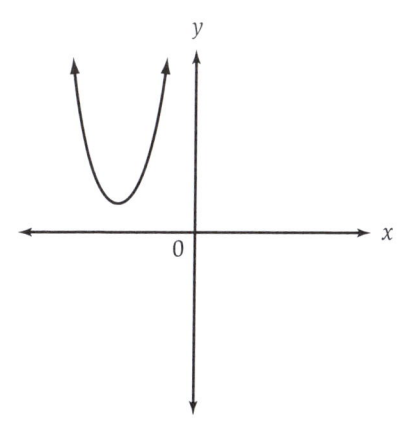

4. Which of the following is the equation of the parabola pictured above?

(A) $y = \frac{4}{5}(x-5)^2 + 2$

(B) $y = \frac{4}{5}(x+5)^2 + 2$

(C) $y = \frac{5}{4}(x-5)^2 - 2$

(D) $y = -\frac{4}{5}(x+5)^2 + 2$

(E) $y = -\frac{5}{4}(x+5)^2 + 2$

| **ESSENTIAL STRATEGIES** |

Here are the two best methods for attacking this item:

1. Ask yourself, Am I familiar with the standard form of the equation for a parabola? If you know it, you're golden.
2. If you don't know it, who cares? Plug in values for x and see what you get.

There's a good chance you won't remember the equation for a parabola, and unfortunately the Math section reference area does not provide this bit of information. So your best bet is to choose a value for x and plug it into the five answer choices. Look at where the parabola dips to its lowest. It's in the part of the graph where x is negative. Now look at your answer choices. What would happen if $x = -5$? If that occurs for choice **A**, you get a big messy number for y. If you plug in $x = -5$ for choice **B**, that whole mess of numbers cancels out since the portion inside the parentheses $(x + 5)$ becomes zero. You're left with $y = 2$, and if you look at the parabola, you could guess that (-5, 2) is the lowest point on the parabola.

So only answer choices with $(x + 5)$ should be considered, meaning you can get rid of **A** and **C**. Now plug in $x = 0$. You should get two answers (choices **D** and **E**) that say y has a negative value when x equals zero, and one choice, **B**, with a positive value. Look at that parabola. It's heading toward the sky, baby! And so y is positive, which means the answer must be **B**.

Note that this method will only work on items where the answer choices are equations or functions. If there is no x in the item, you will not be able to plug in values for it.

TEST-TAKING STRATEGIES

In addition to all the geometry concepts and step methods you have learned, you must also arm yourself with some broad SAT test-taking strategies. If you don't have the correct overall approach to the SAT, all the geometry work you've done will fall by the wayside.

So let's start at a party. Imagine your friends have invited you to a cool house party. Hundreds of people are going to be there, and the party promises to be great fun. (Legal fun, of course.) You had to work that day, so you arrive at the party several hours after it started.

When you walk into the house, do you search for your friends first or do you pick out the scariest-looking person and start talking to him? Unless you like being snubbed, odds are you go through the house the first time looking for friends and other people you know. On a second pass, you might chat with some strangers, but since this is sometimes awkward, you don't usually jump into meeting them at the start.

PACING

Approach every SAT Math section like this party. On the first go-around, look at and answer the easy, familiar items first. This process is made simpler by the fact that every Math section is set up in order of difficulty. The first item is the easiest; the next item is a tiny bit tougher, and so on until the end. A typical twenty-item Math set breaks down the following way:

Difficulty	Item Number
Easy	1–6
Medium	7–14
Hard	15–20

Keep this chart in mind as you take practice tests, but also remember that the order of difficulty is simply based on what the test-makers consider to be easy, medium, and hard. You are your own person, and you may find item 15 to be much easier to complete than item 5.

Answering every item in order—no matter how long it takes—is a classic SAT mistake. Students start with item 1 and then just chug along until time is called. Don't be that chugger! If an item takes more than a minute to solve, skip it and move on to the next item. The goal on the first run-through of an SAT Math section is to answer the items with which you're most comfortable. Save the other items for the second go-around. Although the next item is statistically a little tougher, you might find it easier to answer.

THE SAT IS NOT A NASCAR-SANCTIONED EVENT

The SAT is a timed test, and some people take this to mean that they should answer items as quickly as possible. They cut corners on items to speed through a section. This is a classic error and an especially disastrous policy on SAT Math, since the "easiest" items are at the beginning. If the last item had a fifty-point bonus attached to it, things would be different, but every item counts the same. Getting two hard items right won't do you any good if you've missed two easy items in your haste. You'll be better off answering the easy items correctly and then using whatever time you have left to take an educated guess at the remaining harder items.

Accuracy counts more than speed. First, go through a section and answer all the items that come easily to you. Then:

- Take the time to answer every one of these items *correctly*.
- Take another shot at the remaining items during the second run-through. If you spend two minutes, and the item still doesn't yield an answer, take a guess and move on.

| **TEST-TAKING STRATEGIES** |

To achieve the second point, avoid choosing the answer that looks "right" at first glance. On easy items, this choice may well be the correct answer. For items numbered 12 or higher, an answer that screams, "Ooh! Ooh! Pick me and hurry on!" should be handled like a live snake. SAT distractors are designed to catch the eye of students in a hurry. More often than not, an answer choice for a hard item that looks too good to be true is exactly that: *too good to be true.*

Remember, if you can safely eliminate one of the answer choices as being wrong, you should take a guess because you can beat the wrong-answer penalty.

For most students, the best method for picking up points in the Math section is by:

- Answering all the easy items correctly
- Slowing down and catching most of the medium items
- Getting 25 to 50 percent of the hard items right

This approach is not as thrilling as getting the hardest five items right (while chancing tons of mistakes along the way), but it does put you on the best path to a high score.

WEAR YOUR NO. 2 DOWN TO A NUB

There are many students who are afraid of placing a smudge on their test booklets. These students don't write down any formulas or equations. They don't make any new diagrams. They don't score very well on the SAT either.

Get over your respect for the SAT test booklet. Write all over those rough, recycled pages. When you are finished, your test booklet should be covered with scrawls, notes, computations, and drawings. In fact, the simple act of writing something down for every item will help improve your SAT score. It forces you to put your thoughts down on paper, instead of trying to solve items in your head. If you try to answer items in your head, the SAT will chew you up and spit you out.

Be a smart test-taker. Jot down everything you can.

| **TEST-TAKING STRATEGIES** |

TAKE A DEEP BREATH AND . . .

Don't freak out when you take the SAT. Sure, the test is important, but many people act as if their entire lives depend on how they do on this one exam. It's not true! It's just one test, and you can even take it over again.

On test day, you want to sit down feeling confident and positive. Do what you have to do to get into that mindset—wear a lucky bracelet, do a hundred push-ups, write love poetry—because you need to believe in yourself when taking the SAT. A positive outlook increases your willingness to take an educated guess on a tough geometry item instead of leaving it blank. It helps you trust your inner ear, enabling you to answer a grammar item even though you don't know the exact grammatical rule being tested. A positive approach to the SAT is more important than any single fact or strategy you could learn. Banish anxiety from your mind, and all the skills and strategies you've learned to prepare for the SAT will take its place.

THE 10 MOST COMMON MISTAKES

As you prepare, keep the following common mistakes in mind. Some are mistakes to avoid when taking the actual test. Others are mistakes to avoid during your preparation for the test.

1. Forgetting one of the essential geometry concepts. (Know those rules and formulas!)
2. Forgetting to look for hidden, unstated geometrical figures (especially triangles) that hold the key to an item.
3. Trying to work an item in your head instead of writing your work down.
4. Failing to pick the answer the item asks for. Often this is not the same things as finding the critical value needed to solve the item.
5. Failing to work through our practice sets—*reading* this book is not enough!
6. Failing to practice the step methods on every practice test item. You need these methods when the answer isn't obvious to you.
7. Forgetting to eyeball a Treasure Map item to rule out unlikely answer choices.
8. Refusing to guess when you've eliminated one answer choice.
9. Answering every item in order.
10. Rushing through a set instead of thinking each item through.

CONCLUSION

Without practice, you won't master SAT geometry. You've learned quite a bit since you picked up this little book, but now comes the hard part—*you* have to apply it to testlike items. There are five practice sets at the end of this book: four composed of multiple-choice items and one made up of grid-ins. Here are some tips for getting the most out of these items:

- **Do not time yourself on the first practice set.** When you begin, don't worry about time at all. Take as long as you need to work through each set.
- **Read the explanations for all items, regardless of whether you got them right or wrong.** This is critical—always read *all* the explanations for each set's items. The idea is to develop skills that help you score points as quickly as possible. Most important, scoring a point doesn't mean you got it in the most efficient manner. The overarching goal is to *apply* the methods you've learned. Whether you get all, some, or none of the practice items right doesn't matter.

After the first set, you may want to start paying attention to time. Certainly, by the actual test, give yourself about a minute or so per item.

All the vital information and snazzy strategies you learn in this book won't do a lick of good if you don't use them on the day of the test. Sadly, this happens more often than you might think. Students acquire useful tips, but once the test starts on Saturday morning, all of it goes out the window.

To help ensure that this doesn't happen to you, tackle these five geometry sets *using the skills and strategies you have just learned.* Don't worry about how many you get right or wrong; they're just practice sets. Instead, focus on how well you use the techniques you've learned. When you look at a geometry item, can you tell what method would work best to answer it? If it's a Treasure Map item, how long does it take you to unlock the treasure? If the figures are drawn to scale, which answer choices can you eyeball and cross out?

Don't get frustrated by your progress on the practice sets. Every mistake you make on the practice sets is one that you will avoid on the real

| CONCLUSION |

test. Yes, there'll be some geometry rules you don't know, but learning about these on practice items corrects that deficit. When the real SAT rolls around, you'll have yet another geometry fact in your arsenal that you can employ if needed.

ADDITIONAL ONLINE PRACTICE

Once you're done working through the items and explanations in this book, you can practice further by going online to **testprep.sparknotes.com** and taking full-length SAT tests. These practice tests provide you with instant feedback, delineating all your strengths and weaknesses.

Also, be sure to take the free geometry posttest to see how well you've absorbed the content of this book. For this posttest, go to **testprep.sparknotes.com/powertactics**.

AND FINALLY . . .

The goal of this book is to show you effective methods for answering SAT geometry items. We hope this helps strip away some of the mystery about the SAT that causes so many students to freak out on test day. You should realize that the SAT is not a perfect indicator of your math ability; it simply tests your knowledge on a narrow range of math topics. Master those topics and you will conquer the SAT.

On to the practice items!

THE PRACTICE SETS

PRACTICE SET 1

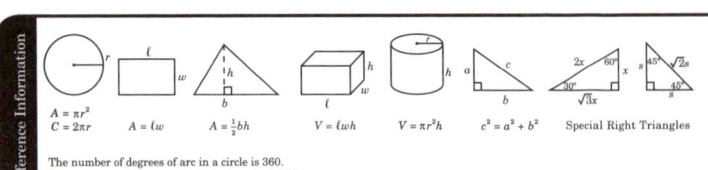

Reference Information

$A = \pi r^2$
$C = 2\pi r$
$A = \ell w$
$A = \frac{1}{2}bh$
$V = \ell wh$
$V = \pi r^2 h$
$c^2 = a^2 + b^2$
Special Right Triangles

The number of degrees of arc in a circle is 360.
The measure of degrees of a straight angle is 180.
The sum of the measures in degrees of the angles of a triangle is 180.

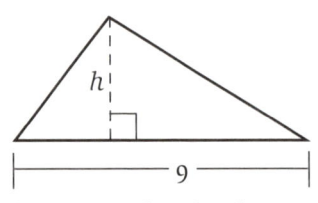

1. If the triangle above has an area of 27, then $h =$

   (A) 3
   (B) 5
   (C) 6
   (D) 8
   (E) 10

| PRACTICE SET 1 |

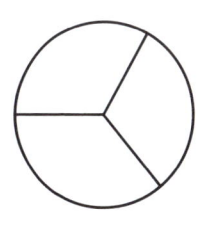

2. The circle above has been divided into three congruent segments. If the circumference of the circle is 12π, what is the area of one segment?

    (A) 4π
    (B) 6π
    (C) 12π
    (D) 24π
    (E) 36π

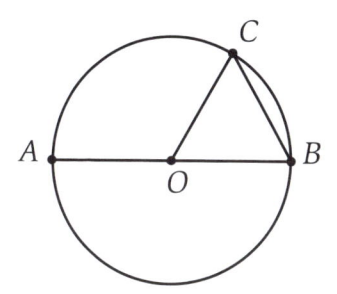

3. If O is the center of the circle above and $\overline{AO} \cong \overline{CB}$, what is the degree measure of angle $\angle AOC$?

    (A) 120
    (B) 115
    (C) 90
    (D) 60
    (E) 45

| PRACTICE SET 1 |

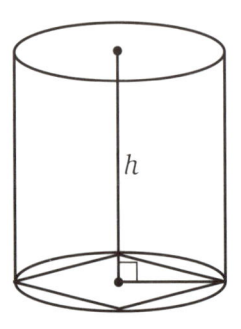

Note: Figure not drawn to scale.

4. A square is inscribed in the base of a right circular cylinder of height 3. If the area of the square is 4, then what is the volume of the cylinder?

   (A) 6π
   (B) $6\pi\sqrt{2}$
   (C) 12π
   (D) $12\pi\sqrt{2}$
   (E) 36π

5. At noon a large marble statue casts a 50-foot shadow. If the angle from the tip of the shadow to the top of the statue is 60°, approximately what is the height of the statue?

The following information may be used to help you find the answer:

$\tan 60° \approx 1.73$
$\sin 60° \approx 0.866$
$\cos 60° \approx 0.5$

   (A) 25 ft
   (B) 43.3 ft
   (C) 50 ft
   (D) 75 ft
   (E) 86.6 ft

ANSWERS & EXPLANATIONS

1. **C**

The first item is usually the easiest, and this one is no exception. All you need to answer the item is the triangle area formula, which you can find

| **PRACTICE SET 1** |

in the Math test-reference section. Now, you might make a careless error if you don't write out your work, so spend the extra ten seconds and guarantee yourself a right answer.

Area of a triangle = (1/2)(base)(height)

$$A = \frac{1}{2}bh$$
$$27 = \frac{1}{2}(9)h$$
$$27 = 4.5h$$
$$\frac{27}{4.5} = \frac{4.5h}{4.5}$$
$$6 = h$$

The correct answer is **C**.

2. **C**

Here is the path you must follow on this item:

Given the circumference → figure out radius → use radius to determine entire area → divide by 3

The circumference is given as 12π, so you can place that into the proper formula to figure out the radius.

$$C = 2\pi r$$
$$12\pi = 2\pi r$$
$$\frac{12\pi}{2\pi} = \frac{2\pi r}{2\pi}$$
$$6 = r$$

Once you have the radius, it's off to the races. In this case, *races* means the "formula for the area of a circle."

$$A = \pi r^2$$
$$A = \pi 6^2$$
$$A = 36\pi$$

Of course, 36π is an answer choice just waiting for those in too much of a hurry to complete the final step. That's the area for the entire circle, but

| PRACTICE SET 1 |

you want the area for one of the three congruent segments. This would be one-third of 36π, or 12π. That's answer choice **C**.

3. **A**

The key clue here is that $\overline{AO} \cong \overline{CB}$. AO is a radius, and so are OC and OB, two legs of the inscribed triangle. All radii are of the same length, and since $\overline{AO} \cong \overline{CB}$, this means that the following four line segments are all equal: AO, OC, CB, and OB. The last three segments form the triangle, and since they are all equal, this means triangle OCB is an equilateral triangle. All angles in an equilateral triangle are 60°, so angle COB = 60°. Angle COB and AOC form a line, so together they equal 180°.

$$\angle COB + \angle AOC = 180$$
$$60 + \angle AOC = 180$$
$$\angle AOC = 120$$

That's choice **A**. If you had no idea how to approach this item, you could have used your eyes. Angle AOC is definitely greater than 90°, and that eliminates choices **C**, **D**, and **E**. At this point, you would have a fifty-fifty chance. Those are good odds on an item you couldn't solve using the step method.

4. **A**

The stem mentions cylinders and squares, but once again it's the unsung triangle that brings home the bacon. The volume formula for the cylinder—one of the useful formulas provided to you at the beginning of the section—is $V = \pi r^2 h$. The height is given to you outright, and by now you are enough of a sleuth to realize that you have to take what's given (the area of the inscribed square) and manipulate this information to get the radius, since that is what is needed for the volume formula.

| PRACTICE SET 1 |

Your journey is aided by the triangle within the inscribed square. Viewing the cylinder from above, it would look like this:

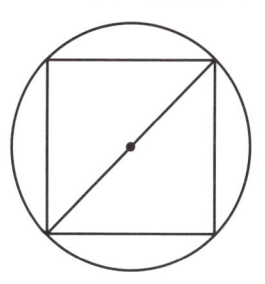

If the area is 4, then the side of the square is 2. Remember that two sides of a square and a diagonal form a 45-45-90 right triangle, which means that the hypotenuse will be the length of a side times $\sqrt{2}$. This makes the diagonal here equal to $2\sqrt{2}$. This diagonal is also a diameter. Since you only need a radius for the volume formula, you must cut it in two to get the radius value of $\sqrt{2}$. Place $\sqrt{2}$ into the volume formula and you get:

$$V = \pi r^2 h$$
$$V = \pi(\sqrt{2})^2 3$$
$$V = \pi(2)3$$
$$V = 6\pi$$

It's choice **A**.

5. **E**

At last, trigonometry rears its ugly head, but it's only there to scare the uninformed test-taker. Rev up your third eye and get ready to draw, because this item needs to be visualized:

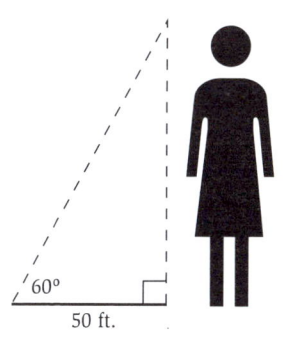

Once drawn, you will see your friend the 30-60-90 right triangle makes an appearance. For these triangles, the side opposite the 60° angle is $\sqrt{3}$

times as large as the side opposite of the 30° angle (the 50-foot shadow, in this instance). Time to use your calculator and determine: $50(\sqrt{3}) \approx 50(1.73) = 86.6$, choice **E**. But wait! We didn't use any trigonometry. Yeah, that was the whole point.

PRACTICE SET 2

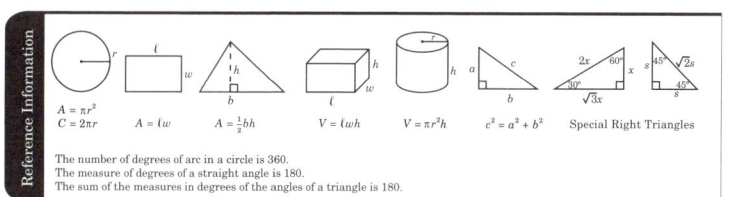

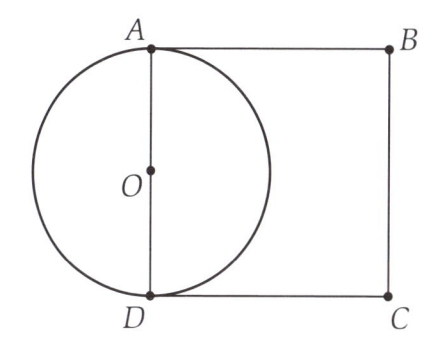

Reference Information

$A = \pi r^2$
$C = 2\pi r$
$A = \ell w$
$A = \frac{1}{2}bh$
$V = \ell w h$
$V = \pi r^2 h$
$c^2 = a^2 + b^2$
Special Right Triangles

The number of degrees of arc in a circle is 360.
The measure of degrees of a straight angle is 180.
The sum of the measures in degrees of the angles of a triangle is 180.

1. If square $ABCD$ has an area of 16, what is the circumference of the circle with center O?

   (A) 2π
   (B) 4π
   (C) 8π
   (D) 16π
   (E) It cannot be determined from the information given.

| PRACTICE SET 2 |

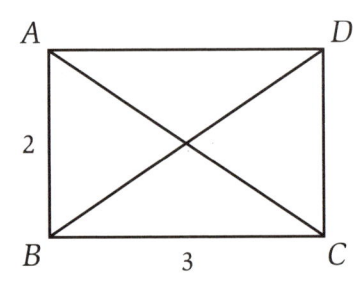

2. If rectangle *ABCD* above has a length of 3 and a height of 2, what is the product of the lengths of the diagonals $\overline{AC}$ and $\overline{BD}$?

   (A) 4
   (B) 6
   (C) 9
   (D) 12
   (E) 13

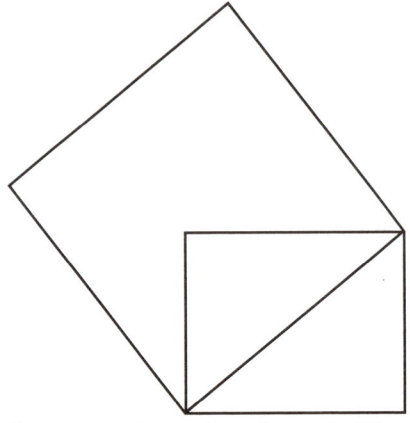

3. One side of the large square is equal to a diagonal of the small square. What is the ratio of the area of the small square to the area of the large square?

   (A) $\frac{1}{4}$

   (B) $\frac{1}{3}$

   (C) $\frac{\sqrt{2}}{3}$

   (D) $\frac{1}{2}$

(E) $\frac{\sqrt{2}}{2}$

4. Point A is at $(-3, 2)$ and point B is at $(3, 10)$. If point D at $(3, 2)$ is the midpoint of line segment AC, what is the perimeter of triangle ABC?

    (A) 36
    (B) 32
    (C) 18
    (D) 16
    (E) 8

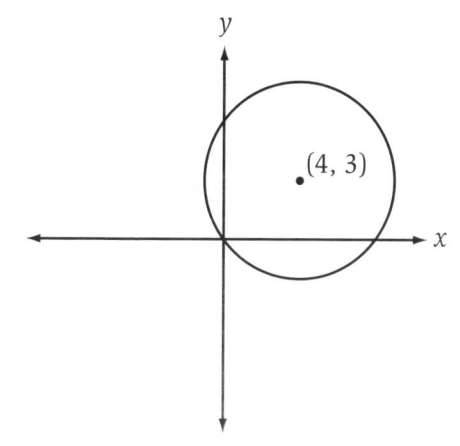

5. The circle above has its center at point $(4, 3)$ and passes through point $(0, 0)$. Which of the following points also lies on the circle?

    (A) $(-1, 1)$
    (B) $(5, -2)$
    (C) $(-2, 4)$
    (D) $(7, 7)$
    (E) $(8, 6)$

ANSWERS & EXPLANATIONS

1. **B**

Here's a standard Treasure Map item. You have two figures, and you need to use the information about one figure (the square) to determine a number associated with the other figure (the circle). Since you have the area of the square, you can first determine the length of each side.

| PRACTICE SET 2 |

$$A = s^2$$
$$16 = s^2$$
$$\sqrt{16} = \sqrt{s^2}$$
$$4 = s$$

Each side of the square is length 4, and one side of the square is the diameter of the circle. This means the diameter of the circle is also 4. Since the circumference of a circle is given by the formula $C = \pi d$, the circumference is $C = \pi d = 4\pi$, answer **B**.

There it is. You've successfully made the jump from information about one figure to another. That was the whole purpose of this item, and you can expect to perform this procedure more than once. Although some items will be more complex, the basic challenge—using the diagram and the information given to hopscotch from one figure to another—stays the same.

2.   **E**

The stem mentions a rectangle, but don't be fooled. This is a triangle item dressed up as a rectangle item. The diagonals are both hypotenuses of right triangles, so the Pythagorean Theorem can be strutted out to determine the length of each diagonal.

$$a^2 + b^2 = c^2$$
$$2^2 + 3^2 = AC^2$$
$$4 + 9 = AC^2$$
$$13 = AC^2$$
$$\sqrt{13} = AC$$

At this point, two things can happen. You can recall that the diagonals of a rectangle are always the same length, or you can run through the Pythagorean Theorem again to figure the length of *BD*. Either way you get $BD = \sqrt{13}$. Since the stem asks for the product of the two diagonals, you now multiply: $(AC)(BC) = (\sqrt{13})(\sqrt{13}) = 13$ and end up with answer **E**.

Always be on the lookout for triangles!

| PRACTICE SET 2 |

3. **D**

There are no numbers given for this item, so plug in some simple ones. You could solve it algebraically, but you could also jab your eye with a pencil: both are *possible*, but neither comes highly recommended.

Make the side of the small square length 2. The area of the small square is 2 times 2, or 4. To find the side of the large square, you can:

1. Unholster the old Pythagorean Theorem, or
2. Realize that two sides of a square and a diagonal form a 45-45-90 right triangle, which means that the hypotenuse will be the length of a side times $\sqrt{2}$

Whichever path you take, the side of the large square is $2\sqrt{2}$. This makes the area of the large square $(2\sqrt{2})(2\sqrt{2}) = 8$. The ratio of the small square over the large square is $\frac{4}{8}$, which reduces to answer $\frac{1}{2}$, choice **D**.

4. **B**

The minute you realized there was no picture with this item, you should have started sketching one yourself. Trying to solve this one mentally could pop your skull. Messy.

Once you place all the points in their proper place, you should get:

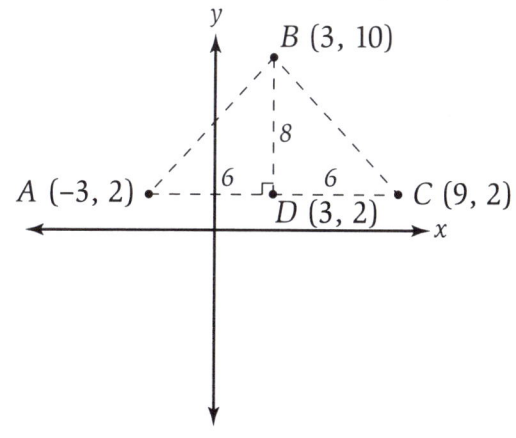

It's a little tough to figure out where C is. If D is the midpoint of AC, and it's six spaces away from A, then it must be six spaces away from C also. That puts C at (9, 2).

You can now determine that AC is equal to 12. Draw a line from B to D and count up. Line segment BD is equal to 8. You now know the value of

| PRACTICE SET 2 |

two sides of a right triangle. You could "Pythagorize" to find *AB* or *BC*, or you could eyeball it and realize that you have two 6:8:10 right triangles, which is just a 3:4:5 right triangle enlarged by a factor of two. Either way both *AB* and *BC* are of length 10. The entire perimeter is 32, choice **B**.

5.   **E**

You can try to eyeball it and figure out which point lies on the circle, but this technique doesn't always work on a hard item. All the points are pretty close, so you could take a guess if you had to, but the odds aren't great for this item.

Determining the radius of the circle may help, so try doing that. Can you guess what hidden figure needs to be drawn? If you said, "Triangle," you're starting to think like a good test-taker. First draw a line from (4, 3) to (4, 0). Now draw the circle's radius. If you draw a radius from (4, 3) to the origin (0, 0), you will have made the hypotenuse of a 3:4:5 right triangle. This means the radius is 5.

Put another way, you can now start at (4, 3) and then go 4 out and 3 up. Each time you do this on the graph, you'll make a radius of 5 and have another point on your circle. If you go 4 out from (4, 3), you'll be at (8, 3). If you go 3 up from there, you'll be at (8, 6), choice **E**.

PRACTICE SET 3

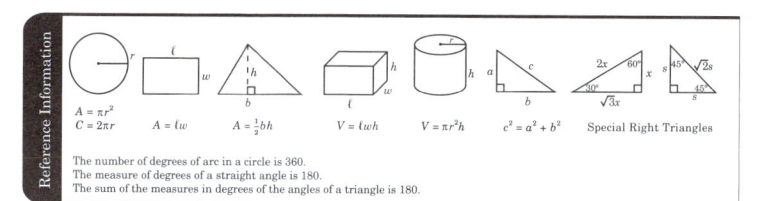

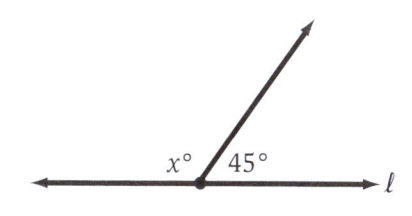

Reference Information

$A = \pi r^2$
$C = 2\pi r$
$A = \ell w$
$A = \frac{1}{2}bh$
$V = \ell w h$
$V = \pi r^2 h$
$c^2 = a^2 + b^2$
Special Right Triangles

The number of degrees of arc in a circle is 360.
The measure of degrees of a straight angle is 180.
The sum of the measures in degrees of the angles of a triangle is 180.

ℓ is a line

1. Using the figure above, which of the following is equal to $2x$?

    (A) 90
    (B) 135
    (C) 180
    (D) 270
    (E) 360

PRACTICE SET 3

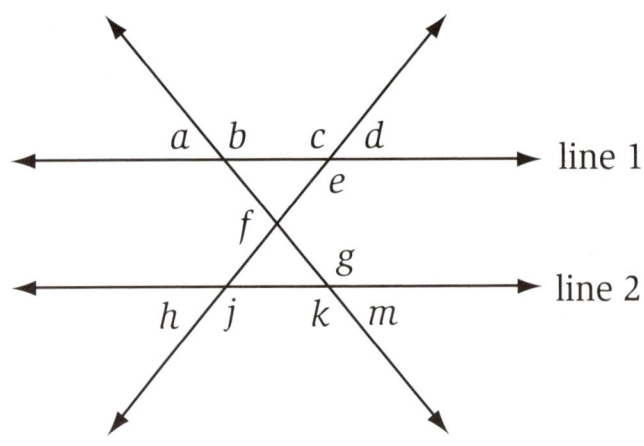

2. In the diagram above, lines 1 and 2 are parallel. Which of the following statements are true?

   I. Angles a and h are equal.
   II. Angles c and g are equal.
   III. Angles k and b are equal.

   (A) I only
   (B) II only
   (C) III only
   (D) I and II
   (E) I and III

3. Let $f(x)$ be defined as $f(x) = -2x + 4$. What is the coordinate of a point that can be found below this function?

   (A) (0, 4)
   (B) (–1, 3)
   (C) (2, 0)
   (D) (2, 4)
   (E) (–1, 5)

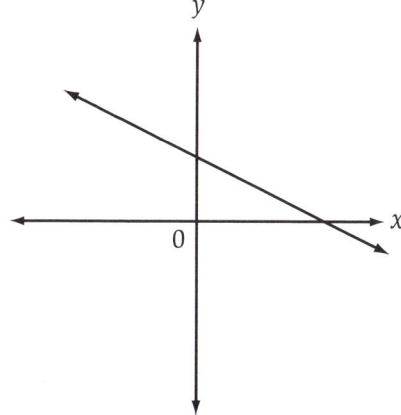

| PRACTICE SET 3 |

4. If the figure above is a graph of the function $f(x) = -\frac{1}{2}x + 3$, which of the following shows the transformation of $f(x)$ into $f(x-4)$?

(A)

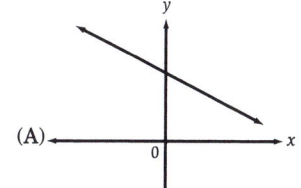

(B)

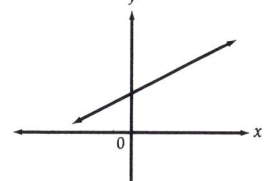

(C)

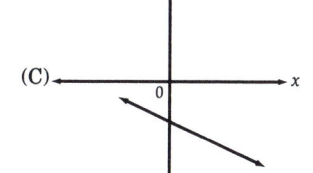

(D)

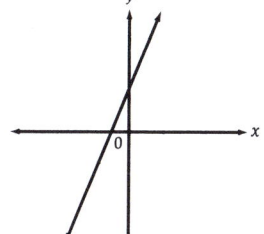

(E)

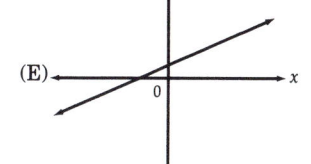

| PRACTICE SET 3 |

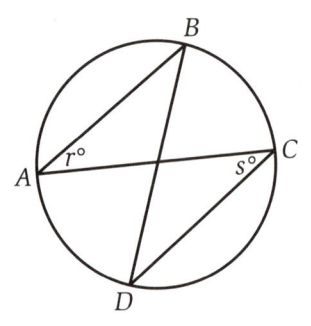

Note: Figure not drawn to scale.

5. The measure of arc AB is 110° and the measure of arc CD is 60°. What is $r° + s°$?

   (A) 85°
   (B) 95°
   (C) 125°
   (D) 170°
   (E) 190°

ANSWERS & EXPLANATIONS

1. **D**

You can eyeball this diagram. The angle x is definitely over 90° in size but less than 180, since that would make it a line. If this is the case, then $2x$ has to be greater than 180° but less than 360°. Selecting from the answer choices, the answer must be **D**.

Ta-da! Now you might be asking, "Why are there such lame distractors on this item?" The answer to this item lies in the method used to solve it. If traveling the conventional route, you would set up an algebraic equation to solve for x based on the fact that there are 180° in a line.

$$x + 45 = 180$$

You would then start manipulating the equation to find x, and after that you would tweak x a bit to find out what $2x$ equals. If you've mistakenly followed this established path, there is a distractor tailor-made to snag you. However by avoiding the established route and tackling the item from another angle, you have avoided the distractors completely.

| PRACTICE SET 3 |

This item was easy, since it was the first item of the set. Applying the "eyeball" strategy makes it even easier.

2. C

Roman Numeral (RN) items are three items for the price of one, since you have to slog through each portion of the item just to get credit for answering one item correctly. To use a highly technical phrase from cost-benefit analysis: that blows.

When you do get around to working the item, always answer one RN and then look to see if any answer choices can be crossed out. For example, let's look at RN I, which claims that angles a and h are equal. There are a bunch of rules regarding similar angles created when a transversal (in other words, a line) crosses two parallel lines, but angles a and h are created by two *different* transversals. You can't just eyeball them and say they look equal; they do look equal, but that could mean that one is 43° while the other is 42°. So RN I is wrong.

Now you can cross out any answer choice that has RN I: choices **A**, **D**, and **E** all fall to our mighty POE ax. The answer must be either **B** or **C**. RN II has two angles that are again created with different transversals, so it's not going to be true either. That leaves **C**, the answer.

There's a mathematical reason why RN III is correct, but why worry about it now? The point is to answer items correctly, which you did by picking **C**. Move on!

3. B

This is a No Map item, so the first thing you should do is sketch out the function. The slope is −2 and the y-intercept is 4, so the line will look like:

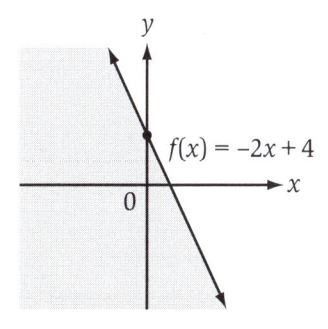

$f(x) = -2x + 4$

Pretty much any point you pick that is below this line will work, but you have to choose from the five answer choices provided. Choices **A** (the y-

| **PRACTICE SET 3** |

intercept) and **C** (the *x*-intercept) are classic distractors because they are values that are easy to spot when you draw out the function. If you're in a hurry, you might select one. Instead, take the time to plot the five answer choices on your graph. You'll quickly see that **B** is the correct answer.

4. **A**

This is the kind of difficult-looking item that gives many people the willies. If you keep your cool, you'll find the item is not too difficult.

All you need to do is follow the instructions. Remember to think like a robot. It helps. You're given an initial equation $f(x) = -\frac{1}{2}x + 3$, and then asked how it will be different once you transform it to $f(x - 4)$. Ignore the funky $f(x)$'s and just replace the x in the first equation with $(x - 4)$. This gives you:

$$f(x) = -\frac{1}{2}x + 3$$
$$f(x) = -\frac{1}{2}(x - 4) + 3$$
$$f(x) = -\frac{1}{2}x + 2 + 3$$
$$f(x) = -\frac{1}{2}x + 5$$

That's the new line you're looking for. As you can see, the slope, $-\frac{1}{2}$, has not changed, so you can eliminate choices **B**, **D**, and **E** since they all feature new slopes. Choice **A** has a *y*-intercept at 5, so it's the right answer.

If the discussion about slopes and intercepts has you shaking your head, review those pages again (pp. 29–30).

5. **B**

The sum of two variables is needed here, which may create problems for some students who incorrectly believe they have to find exact values for the variables. This isn't true! This item requires remembering this obscure formula:

> An arc defined by an inscribed angle is always equal to twice the measure of that angle.

If you can pull this lovely piece of geometrical knowledge out of your hat, you're in business. An entire circle measures 360°, and you are given two arc measures: arc *AB* is 110° and arc *CD* is 60°. These are *not* the arcs opposite *r* and *s*, so a little subtraction will give you the measure of the arcs opposite your two mystery variables.

$$\text{Whole Circle} = 360°$$
$$\text{arc } s + \text{arc } r + \text{arc } AB + \text{arc } CD = 360°$$
$$\text{arc } s + \text{arc } r + 110 + 60 = 360°$$
$$\text{arc } s + \text{arc } r + 170 = 360°$$
$$\text{arc } s + \text{arc } r = 190°$$

The numeral 190 is a good distractor since it's the arc measure for the two variables. However, the item wants the angle measurements, and since the arc measure is twice the angle measure, you must divide 190 by 2 to get 95, choice **B**.

PRACTICE SET 4

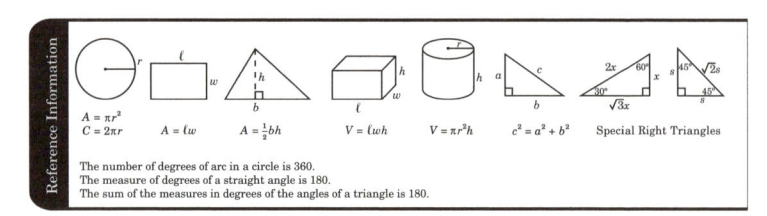

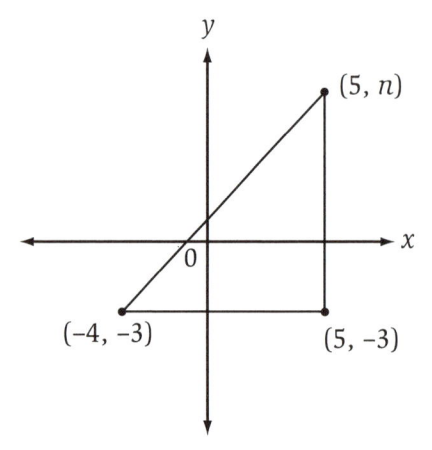

1. The points labeled in the above figure are the vertices of a triangle with an area of 45. What is n?

    (A) 5
    (B) 6
    (C) 7
    (D) 9
    (E) 10

| PRACTICE SET 4 |

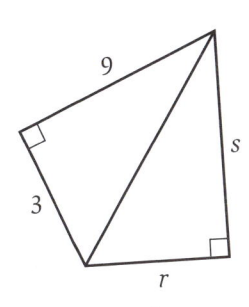

Note: Figure not drawn to scale.

2. What is the value of $r^2 + s^2$?

(A) 27
(B) $3\sqrt{10}$
(C) 90
(D) 243
(E) 729

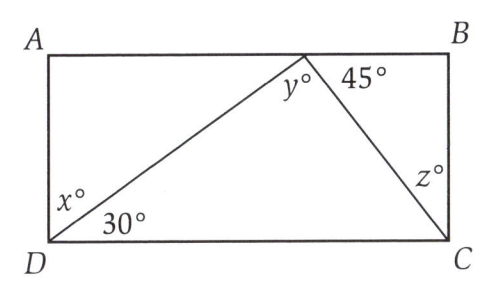

3. If $ABCD$ is a rectangle, what is $x + y + z$?

(A) 75
(B) 160
(C) 175
(D) 180
(E) 210

| PRACTICE SET 4 |

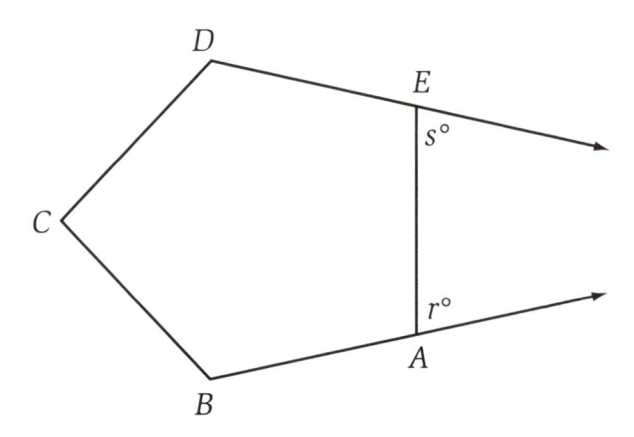

Note: Figure not drawn to scale.

4. If ∠ABC + ∠BCD + ∠CDE = 330°, then $r + s =$

    (A) 150°
    (B) 165°
    (C) 180°
    (D) 195°
    (E) 210°

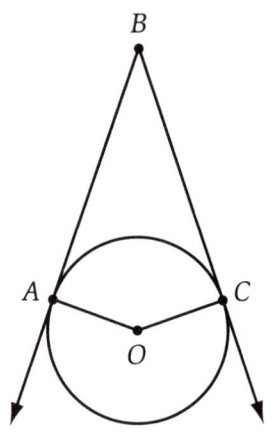

5. Rays $\overrightarrow{BA}$ and $\overrightarrow{BC}$ are each tangent to the circle with center O. If the radius of the circle is 1 and the measure of angle ∠ABC is 60°, what is the area of quadrilateral $OABC$?

    (A) $\frac{\sqrt{3}}{4}$

(B) $\frac{1}{2}$

(C) $\frac{\sqrt{3}}{2}$

(D) $\sqrt{3}$

(E) $2\sqrt{3}$

ANSWERS & EXPLANATIONS

1. **C**

This item is a little harder than the standard "find the area of a triangle" item for the following reasons:

- The introduction of a grid
- No lengths
- The appearance of the variable n as part of an (x, y) point

Sure, it's a little different, but you aren't about to let that throw you. Remember this item when you take the real SAT. You will encounter a triangle formula item similar to this one. Use the step methods on these practice items, and you'll be more than ready for the real thing.

You have the area of the triangle, 45, and you know the base is the distance between $(-4, -3)$ and $(5, -3)$. This is a length of 9, since $|-4 - 5| = 9$. Put these values into the area formula:

$$A = \frac{1}{2}bh$$
$$45 = \frac{1}{2}(9)h$$
$$45 = 4.5h$$
$$\frac{45}{4.5} = \frac{4.5h}{4.5}$$
$$10 = h$$

So the height is 10. Therefore, point $(5, n)$ is 10 spaces above point $(5, -3)$, making $n = 7$ since $|7 - -3| = 10$. That's choice **C**.

Those straight-line bars denote absolute value. There is a long-winded definition of absolute value, but for our purposes all you need to

| PRACTICE SET 4 |

know is that the absolute value of a number is always positive. That's why $|-4 - 5|$ above is equal to 9 instead of –9.

2. **C**

The diagram shows two triangles, or two triangular faces of a pyramid. Just looking at the picture, it seems impossible to find the exact values of r and s since there's not enough information given. Yet the item doesn't ask what r and s are. It asks for $r^2 + s^2$. Look at that expression and think triangles. Can you see what the missing link is? We'll give you some more hints if you can't:

- Think "right triangles."
- Think of a theorem regarding right triangles that would use the expression $r^2 + s^2$.
- It starts with *Pyth* . . .

We hope that your new SAT-savvy mind sees the right triangles, the expression $r^2 + s^2$, and makes the leap to, "I can solve this triangle item using the Pythagorean Theorem." And you can, since the two right triangles share a hypotenuse.

$$r^2 + s^2 = (\text{hypotenuse})^2 = 3^2 + 9^2$$
$$(\text{hypotenuse})^2 = 9 + 81$$
$$(\text{hypotenuse})^2 = 90$$
$$r^2 + s^2 = 90$$

Choice **C** is correct.

3. **E**

Triangles ahoy! Unpack the rule that states the sum of all three interior angles equals 180°, and go to town with it on this item. To start the festivities, you can look at the bottom left corner and realize that angle x and 30° combine to form a 90° right angle. That means $x = 60°$. Now to solve for y. If $x = 60°$ and the angle up by A is equal to 90°, then the third angle in that triangle is equal to 30°. The 30° angle along with angle y and the 45° angle on the other side of y make up a line. Now do some basic math:

$$30 + y + 45 = 180$$
$$75 + y = 180$$
$$y = 105$$

| PRACTICE SET 4 |

Beautiful. Now we need to solve for angle z, which is part of another right triangle. We already have two angles in that triangle, the 45° angle given to us in the diagram and the 90° angle up by B. Surprise, surprise! We have a 45-45-90 right triangle. Angle z = 45°.

Once you have values for the three variables, it's easy enough to sum them up and get: $x + y + z = 60 + 105 + 45 = 210$, choice **E**.

4. **A**

The item only tells you that three of the interior angles equal 330°. Since the figure is not drawn to scale, you have to pull out the formula for the sum of the interior angles of a polygon. This formula is $(n - 2)180°$, where n is the number of sides. For our pentagon, the interior angles must equal $(5 - 2)180° = (3)(180°) = 540°$.

So this means:

$$\angle ABC + \angle BCD + \angle CDE + \angle DEA + \angle EAB = 540$$
$$330 + \angle DEA + \angle EAB = 540$$
$$\angle DEA + \angle EAB = 210$$

How do these two angles relate to r and s? One of them forms a line with r, and the other forms a line with s. A line is 180°, so you can write:

$$\angle EAB + r = 180 \text{ and } \angle DEA + s = 180, \text{ so}$$
$$\angle EAB + r + \angle DEA + s = 180 + 180$$
$$\angle EAB + \angle DEA + r + s = 360$$
$$210 + r + s = 360$$
$$r + s + 150$$

There's your answer, choice **A**. The item makes you work for it, but that's why it comes later in the set.

5. **D**

Draw the line *OB*. It gives you two triangles. Since tangential lines are perpendicular to a circle, you have two right triangles, *OAB* and *OCB*. The next step shows why this item is the hardest in the set. You have to realize that *OB* bisected angle *ABC*, creating two equal 30° angles.

You now have two identical 30-60-90 right triangles with quadrilateral *OABC*. The length of the shortest side in each triangle is 1, since this side is also the radius of the circle.

Go to town! Start crunching numbers! Because we're dealing with a 30-60-90 right triangle, we know that *BC* is $\sqrt{3}$ (remember the ratio between sides in these special right triangles). *BC* is also the height of our triangle, so the area of the triangle is:

$$A = \frac{1}{2}bh$$
$$A = \frac{1}{2}(\sqrt{3})1$$
$$A = \frac{\sqrt{3}}{2}$$

Now multiply this by 2 to find the area of quadrilateral *OABC*:

$$2\left(\frac{\sqrt{3}}{2}\right) = \sqrt{3}$$

That's choice **D**.

PRACTICE SET 5

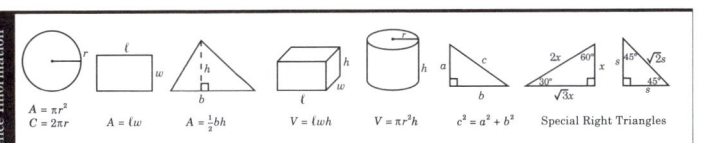

Reference Information

$A = \pi r^2$
$C = 2\pi r$

$A = \ell w$

$A = \frac{1}{2}bh$

$V = \ell w h$

$V = \pi r^2 h$

$c^2 = a^2 + b^2$

Special Right Triangles

The number of degrees of arc in a circle is 360.
The measure of degrees of a straight angle is 180.
The sum of the measures in degrees of the angles of a triangle is 180.

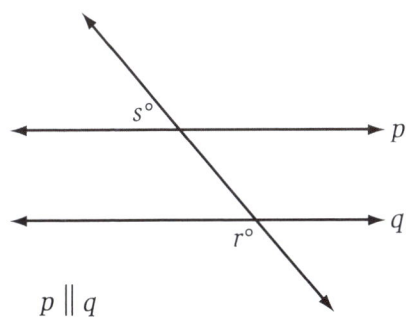

$p \parallel q$

1. If $t = 2s + 15$, what is t?

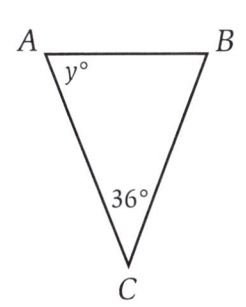

2. If $AC = BC$, then $y = $?

3. The sum of the interior angles of a regular polygon equals 1260°. How many sides does the polygon have?

4. Two lines pass through point (2, 4). Line 1 also passes through (−2, 1), while line 2 passes through (4, 0). What is the slope of one of these lines?

| PRACTICE SET 5 |

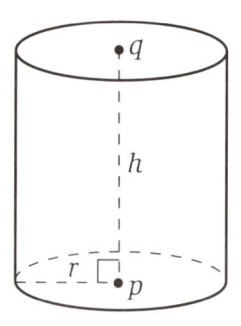

5. The volume of the right circular cylinder pictured above is 81π. If the height of the cylinder is three times the radius, what is the diameter of circle P?

ANSWERS & EXPLANATIONS

1. **125**

Since lines p and q are parallel, angles s and t are supplementary angles. This means they are equal to 180.

$$s + t = 180$$
$$s + 2s + 15 = 180$$
$$3s + 15 = 180$$
$$3s = 165$$
$$s = 55$$

Keep in mind the item asks for the value of t, not s. Placing this value of s into $t = 2s + 15$ gives you:

$$t = 2s + 15$$
$$t = 2(55) + 15$$
$$t = 125$$

2. **72**

If $AC = BC$, then triangle ABC is an isosceles triangle. The angles opposite the equal sides are also equal, so $\angle ABC = \angle BAC$. $\angle BAC$ is just another name for angle y. Since the sum of the interior angles of a triangle is 180, set up the equation like this:

| PRACTICE SET 5 |

$$y + \angle ABC + 36 = 180$$
$$y + y + 36 = 180$$
$$2y = 144$$
$$y = 72$$

3. **9**

The key to this item lies in remembering that the formula for the interior angles of a regular polygon is $(n-2)180$, where n is the number of sides for that polygon.

$$(n-2)180 = 1260$$
$$\frac{(n-2)180}{180} = \frac{1260}{180}$$
$$n - 2 = 7$$
$$n = 9$$

It's a nine-sided polygon.

4. **3/4**

This is a No Map item. Since there's no diagram, you should draw one.

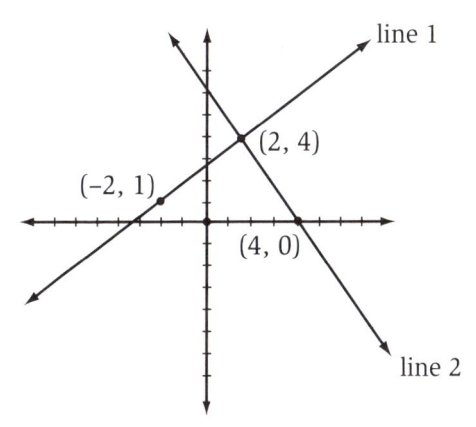

A quick sketch will help you see that line 2 has a negative slope. Since you can't grid in negative values, you have to find the slope of line 1. This line passes through points (2, 4) and (–2, 1), so placing these two points into the slope formula gives you:

PRACTICE SET 5

$$m = \frac{y_1 + y_2}{x_1 - x_2} = \frac{4-1}{2--2} = \frac{3}{4}$$

5. 6

You're given the volume of the cylinder, and since that's the only number you have to work with, it's the best place to start. The formula for the volume of a right circular cylinder is:

$$V = \pi r^2 h$$
$$81\pi = \pi r^2 h$$
$$81 = r^2 h$$

This might seem like the end of the road, but the item gives you one more fact: the height of the cylinder is three times the radius. A light might flash in your head at this point, but if not, just create an equation that shows this relationship:

$$h = 3r$$

Now place this into the formula above:

$$81 = r^2 h$$
$$81 = r^2(3r)$$
$$81 = 3r^3$$
$$27 = r^3$$
$$3 = r$$

At this point, you might breathe a sigh of relief and then glibly place the number 3 in the grid. Don't! The item asks for the *diameter*, so the correct answer is twice the radius, 6.

SPARKNOTES
Power Tactics for the New SAT

The Critical Reading Section
Reading Passages
Sentence Completions

The Math Section
Algebra
Data Analysis, Statistics & Probability
Geometry
Numbers & Operations

The Writing Section
The Essay
Multiple-Choice Questions: Identifying Sentence Errors, Improving Sentences, Improving Paragraphs

The New SAT
Test-Taking Strategies
Vocabulary Builder

SPARKCHARTS exclusively at Barnes & Noble

lost your class notes?

Don't worry, it's all here in your binder.

SPARKCHARTS™ are the perfect studying companion and reference tool for anyone in high school, college, grad school, or even the office! These sturdy, laminated outlines and summaries cover key points, with diagrams and tables that make difficult concepts easier to digest. So put the antacid away and head to Barnes & Noble!

Smarter, better, faster™ in over 100 subjects! ✹ Algebra I ✹ Algebra II ✹ Anatomy ✹ Th Bible ✹ Biology ✹ Chemistry ✹ English Composition ✹ English Grammar ✹ English Vocabular ✹ Essays and Term Papers ✹ Literary Terms ✹ Microsoft Excel ✹ Geometry ✹ Math Basics ✹ Spanish Medical Terminology ✹ Periodic Table ✹ Physics ✹ The SAT ✹ Spanish Grammar ✹ Spanish Verbs ✹ Spanish Vocabulary ✹ Study Tactics ✹ U.S. History ✹ World Map ✹ ...and many more

http://www.sparknotes.com/buy/charts

AS SEEN ON
WWW

Don't bother
...we've done it for you!

EXCLUSIVE TO **Barnes & Noble**

la felicidad

Andrew Johnson

Sick of scribbling on index cards?

Making that bump on your finger worse?

Planck's cons

Relax... SparkNotes has it covered!

aerobi

SparkNotes®
Study Cards

not only come in the most popular subjects, but they also translate difficult subject matter into digestible tidbits, making them the must-have studying companion. And if you still want to write your own cards, we've got blank ones too!

Titles include
Biology
Blank Cards
Chemistry
English Grammar
English Vocabulary
French Grammar
French Vocabulary
Geometry
Organic Chemistry
Physics
Pre-Algebra
Spanish Grammar
Spanish Vocabulary
U.S. History

SAT vocabulary novels

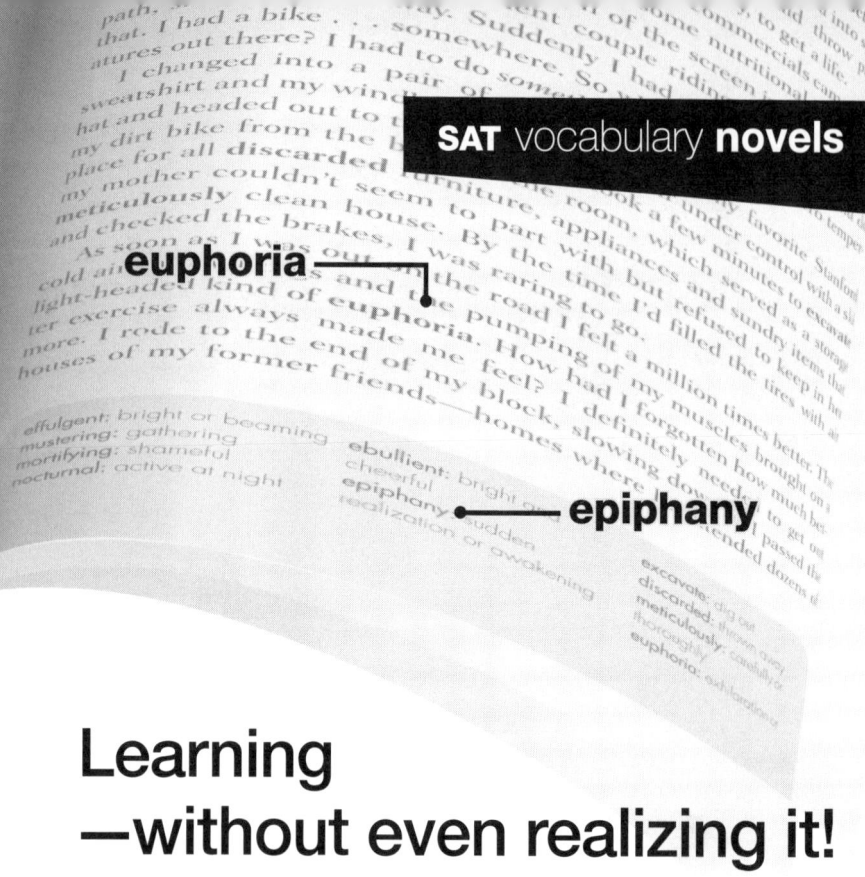

Learning
—without even realizing it!

Need to study for the SATs, but would rather read a good book? Now SparkNotes® lets you do both. Our **SAT Vocabulary Novels** are compelling full-length novels wtih edgy and mature themes that you'll like (honest). Each book highlights more than 1,000 vocabulary words you'll find on the SAT. Brief definitions appear at the bottom of the page, so you can sit back, read a good book, and get some serious studying done—all at the same time!

excerpts and more at www.sparknotes.com/**buy/satfiction/**

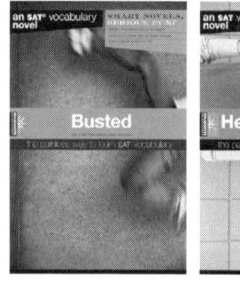

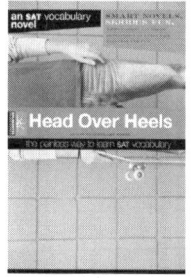

Busted | Head Over Heels | Sunkissed | Vampire Dreams